A handbook of terms used in algebra and analysis

A handbook of terms used in algebra and analysis

compiled by **A. G. Howson**

Cambridge
at the University Press 1972

CAMBRIDGE UNIVERSITY PRESS
Cambridge, New York, Melbourne, Madrid, Cape Town, Singapore, São Paulo, Delhi

Cambridge University Press
The Edinburgh Building, Cambridge CB2 8RU, UK

Published in the United States of America by Cambridge University Press, New York

www.cambridge.org
Information on this title: www.cambridge.org/9780521084345

First published 1972
Re-issued in this digitally printed version 2008

A catalogue record for this publication is available from the British Library

Library of Congress Catalogue Card Number: 71-178281

ISBN 978-0-521-08434-5 hardback
ISBN 978-0-521-09695-9 paperback

Contents

Preface

The enormous increase in mathematical activity and knowledge during the past half century has not been achieved without a corresponding increase in the number of terms which mathematicians use. Not only have names had to be given to new concepts and objects, for example categories and functors, but there has also been a need to attach new and/or more precise meanings to certain terms such as 'function' which have been used, in some sense or other, for centuries. Again, deeper insight into mathematical structure has often yielded an alternative way of approaching such well-established ideas as that of a derived function. This creation of new definitions and rewriting of old ones has not made life any the easier for the reader of mathematics. Certainly, an explanation of any of the terms can be found somewhere, but finding the correct source can be time consuming. The attractions of a reference book of definitions are, therefore, obvious.

Alas, the difficulties of compiling such a work are no less obvious! As soon as such a book were to appear it would be months out of date, for each issue of every mathematical journal can be expected to introduce at least one new term to the vocabulary of mathematics and frequently a new symbol – or an alternative usage of an old one – to accompany it. Moreover, talents equal to those of the troops of Bourbaki would be required to produce a comprehensive and authoritative work.

The objectives of this handbook, then, must be somewhat circumscribed. It is, for example, intended to meet the needs of the undergraduate and the school teacher rather than the university lecturer or the research student. It is concerned only with algebra and analysis. Nevertheless, I hope that within these limits the handbook will prove of value and that by its arrangement into sections, its examples and its notes, it will give some indication of, and some feeling for, the mathematics that has given rise to the definitions listed. Some cynics have asserted that 'modern mathematics' is 'all definitions', and it cannot be denied that it does contain a great number of them. It must be stressed, therefore, that the terms defined here are only the 'words' of mathematics and that a mathematician is interested not only in learning new 'words' but in 'creative writing' and in appreciating the subject's 'literary heritage'.

Presenting a definition out of context is not a simple matter. 'How do I get to Newcastle from here?' is a question which has many answers. Is the questioner travelling on foot or on horseback, by bicycle or by car, or, come to that, by boat? Does he want the fastest route or the least congested? Would he like to see the new motorway, or travel on minor roads past some fascinating historical monuments? These questions all have their mathematical analogues – as does the apocryphal Irishman's answer that 'If I wanted to go to Newcastle, I shouldn't start from here.' There is always the chance, too, that the questioner really wanted to go to Newcastle-under-Lyme and has been directed in error to Newcastle-upon-Tyne. In an attempt to meet some of the analogous mathematical problems, many terms have been defined in this handbook in alternative ways; for example, 'continuity' is discussed in three different sections, namely, Metric spaces, Topological spaces, and Real-valued functions of a real variable. Theorems too have often been stated at alternative levels of generality. I have also tried to draw attention to words and phrases which are used in different senses by different authors, for example 'ring' and 'set of natural numbers'.

It would, of course, be difficult to answer the traveller without using numbers, yet no one is likely to preface his reply by an account of how the number system is constructed. In a similar manner, one cannot define terms in mathematics without using some words or symbols taken from logic. For this reason, the first section of the handbook describes those words and symbols which we shall wish to borrow from logic, without, however, concerning itself with the foundations of that subject.

The use of symbols in mathematical literature is, if anything, even more bewildering than that of defined terms. Only the '+' sign in the context of the addition of real numbers springs to mind as a symbol used by all writers in an unambiguous manner. (Lest readers press the case of the companion '−' sign, I was informed by a Scottish Inspector of Schools that he visited one primary school where the teacher taught the class that '−' was to be translated as 'from' and that the children were, therefore, to write equations such as $2-8 = 6$. The teacher explained to the inspector that the children experienced less difficulty with this convention!) I have tried, therefore, when introducing symbols to list any alternatives which are widely used.

In particular, in the sections on differentiation I have presented definitions, results and examples in a variety of notations – both 'ancient and modern' – in the hope that the opportunities thus provided for comparing the different notations will more than compensate for any possible loss in clarity.

Whilst producing this book I have received help and encouragement from many sources. As far as content is concerned, I am indebted to a host of mathematicians ranged alphabetically from Abel to Zorn and chronologically from Pythagoras to Cohen! Several colleagues or former colleagues at Southampton have been kind enough to read sections of the book and to offer most valuable advice. In particular, I should like to acknowledge my especial indebtedness to Professor T. A. A. Broadbent and to Dr Keith Hirst who read the final manuscript and whose observations led to the removal of several errors and, I hope, to a more readable book. I am also most grateful to Jennifer, my wife, for – amongst many other things – her assistance in the compilation of the index to this book.

May, 1971 A. G. HOWSON

1 Some mathematical language

Connectives

Mathematics is concerned with statements or propositions of a particular kind, namely those which are composed of mathematical signs and objects. Such statements are often called **relations.**

From these statements we can build up others by means of certain logical signs known as **connectives.** The basic connectives which we use to construct compound statements are

not	denoted by	$\sim$ ($\neg$),
or	denoted by	$\vee$,
and	denoted by	$\wedge$ (&),
if...then	denoted by	$\Rightarrow$ ($\rightarrow$, $\supset$),
if and only if	denoted by	$\Leftrightarrow$ ($\leftrightarrow$, $\equiv$).

These connectives are used in everyday speech, but often in an ambiguous and confusing manner. It is necessary then to fix their meaning more precisely and this is done by means of **truth tables** which provide information regarding the truth or falsehood of a compound statement in terms of the assumed truth or falsehood of its constituent parts. Thus the operation of **negation** (not) is defined by the table

X	$(\sim X)$
T	F
F	T

which indicates that if the statement X is true then its negation, $(\sim X)$, is false, and if the statement is false then its negation is true.

Conjunction (and), **disjunction** (or), **logical implication** (if... then), and **logical equivalence** (if and only if) are similarly defined by

X	Y	$(X \wedge Y)$	$(X \vee Y)$	$(X \Rightarrow Y)$	$(X \Leftrightarrow Y)$
T	T	T	T	T	T
T	F	F	T	F	F
F	T	F	T	T	F
F	F	F	F	T	T

Note. (i) In this interpretation of 'or' we say that $(X \vee Y)$ is true when X is true *and* Y is true. This is sometimes known as the '**inclusive or**', and, it will be noted, is not always the everyday usage of 'or'.

(ii) In the statement $(X \Rightarrow Y)$ (sometimes known as a **conditional** statement), X is known as the **antecedent** or **hypothesis** and Y as the **consequent** or **conclusion.**

Just as in everyday speech it is possible to express the same argument in different ways, it can happen that two mathematical statements 'say the same thing'. We say that two compound statements are **equivalent** when they have identical truth tables. Thus, for example, $(X \Rightarrow Y)$ and $((\sim X) \vee Y)$ are equivalent since we have

X	Y	$(\sim X)$	$X \Rightarrow Y$	$((\sim X) \vee Y)$
T	T	F	T	T
T	F	F	F	F
F	T	T	T	T
F	F	T	T	T

Note. (i) It follows that $\Rightarrow$ can be expressed in terms of $\sim$ and $\vee$, as indeed can all the other connectives. (The connectives can also be expressed in terms of $\sim$ and $\wedge$.)
More basic still are the **Sheffer stroke** $|$ and the connective $\downarrow$ defined by

X	Y	$(X \mid Y)$	$(X \downarrow Y)$
T	T	F	F
T	F	T	F
F	T	T	F
F	F	T	T

since all the other connectives can be defined in terms of $|$ (or $\downarrow$) alone.

(ii) The fact that two statements are equivalent is often used to simplify the proofs of theorems. Thus, to prove that A is true, we show that B, a statement equivalent to A, is true.

Axioms

Certain statements (relations) play a particularly important role in any mathematical system. These are relations which are stated explicitly, once and for all, and which are then known as the **axioms** of the system. Relative to the system, all axioms are said to be **true.**

In addition to this type of axiom there is a second type, the **logical axioms**, which tell us how the connectives are to be manipulated – these are the rules of reasoning.

We now say that a statement (relation) in a particular system is **true** (i.e. true relative to the system), if it can be obtained by repeated application of the rules:

(*a*) every relation obtained by applying an axiom is true,

(*b*) if R and S are relations such that $(R \Rightarrow S)$ is true and if, in addition, the relation R is true, then the relation S is true.

A relation R is **false** if $(\sim R)$ is true.

Note. (i) (*b*) is known in logic as the rule of **modus ponens.**

(ii) Given a relation $(R \Rightarrow S)$ we define its **converse** to be $(S \Rightarrow R)$, its **inverse** to be $((\sim R) \Rightarrow (\sim S))$, and its **contrapositive** to be $((\sim S) \Rightarrow (\sim R))$. A frequently used rule of reasoning is that

$$((R \Rightarrow S) \Leftrightarrow ((\sim S) \Rightarrow (\sim R))) \quad \text{is true,}$$

i.e. that a conditional relation and its contrapositive are equivalent (and, hence, that the converse relation is equivalent to the inverse relation).

(iii) It is not the case that a relation must be true or false. There are relations which are **undecidable** in the sense that they can be neither proved nor disproved. (Here a mathematical system differs from the propositional calculus of logic.)

Variables and quantifiers

The relations which we consider in mathematics frequently contain letters representing indeterminate mathematical objects which we refer to as **variables.** If R contains the variable x, we can replace x wherever it occurs in R by a mathematical object A, e.g. a number or a set. When we do this the resulting assembly of letters and signs is again a relation and is known as the relation obtained by **substituting** A for x in R. A is said to **satisfy** the relation R if the relation so obtained is true (thus 3 satisfies the relation $x^2+2 = 11$).

To indicate that x occurs in the relation R we write $R\{x\}$ and we denote the relation obtained by substituting A for x by $R\{A\}$. A relation of the form $R\{x\}$ is often known as a **predicate.**

Two questions which are frequently asked of mathematical relations containing a variable are: 'Is the relation true *for every A* which we care to substitute for x?' (in practice it would be assumed that A would belong to a particular class of objects, for otherwise the relation obtained by substituting for x would frequently be meaningless, e.g. $(\text{triangle})^2+2 = 11$) and 'Is there *some A* (belonging to a particular class) for which the relation is true?' These two notions lead to the introduction of the two **quantifiers** $\forall$ and $\exists$.

In mathematics we use as a shorthand for the phrases

for all, for each, for every, for any, given any

the sign $\forall$ which is known as the **universal quantifier.**

Thus we write

$$(\forall x) \quad (x \text{ is a real number} \Rightarrow (x+2)^2 = x^2+4x+4)$$

to indicate that the relation

$$(x+2)^2 = x^2+4x+4$$

is satisfied by every real number x.

Similarly, we express the fact that *there exists* an object which satisfies the relation $x^2+2 = 11$ by writing

$$(\exists x) \quad (x^2+2 = 11),$$

where $\exists$, the **existential quantifier**, is interpreted as shorthand for '*there exists... such that*' or '*for some*'.

Note. (i) The usage of $\forall$ and $\exists$ as explained above corresponds to the everyday usage of the logical connectives. Their usage in logic, from whence the symbols are borrowed, is, of course, more circumscribed.

(ii) It is often convenient to work with restricted notions of quantification and to write

$$(\forall x)\ ((x+2)^2 = x^2+4x+4)$$

rather than $(\forall x)$ (x is a real number $\Rightarrow$ $(x+2)^2 = x^2+4x+4$) when it is clearly understood from the context that x is a real number.

(iii) Some authors choose to define $\forall$ in terms of $\exists$ and $\sim$ since $(\forall x)\ (R\{x\})$ and $\sim ((\exists x)\ (\sim R\{x\}))$ are equivalent statements.

(iv) A useful formula which enables one to write down the negation of a relation containing two or more quantifiers relating to different variables is $(\sim ((\forall x)\ (\exists y)\ R\{x, y\}))$ is equivalent to $((\exists x)\ (\forall y)\ (\sim R\{x, y\}))$.

(v) The symbol $\exists! x$ is often used to denote 'there exists a unique x,...'.

When a variable occurs in a relation which is quantified, then we say that x is a **bound variable** in that relation, e.g. x is bound in the relation $(\exists x)\ (x^2+2 = 11)$. If no quantifier appears in the relation then x is said to be a **free variable** or an **unknown** and accordingly we speak of $R\{x\}$ as defined above as a **predicate with free variable** x.

A relation containing only bound variables or no variables at all is said to be **closed** whilst a relation containing a free variable is said to be **open.** Closed statements have the property of being true, false or undecidable.

Examples. (i) $(\exists x)\ (x^2+2 = 11)$ is true;

(ii) $(\exists x)\ (x^2 = -1)$ is false (assuming that we are again using quantification restricted to the real numbers; if we do not make this assumption then the relation is true);

(iii) (x is a prime) is an *open* relation which is neither true nor false in itself but only becomes so when a particular object is substituted for x. Those objects of a given set which satisfy an open statement form the **solution set** of the statement relative to that set. Thus the *solution set* of $x^2 = 2$ relative to the real numbers is the set $\{+\sqrt{2}, -\sqrt{2}\}$. Relative to the rational numbers (p. 24), the *solution set* of $x^2 = 2$ is empty (p. 9).

Equality

The mathematical sign of **equality** is introduced in order to form relations of the type $a = b$ which we take intuitively to mean that the objects a and b are identical.

The rules which we require for manipulating equalities are:

(*a*) $(\forall x)\,(x = x)$,

(*b*) $(\forall x, y)\,((x = y) \Leftrightarrow (y = x))$,

(*c*) $(\forall x, y, z)\,(((x = y) \wedge (y = z)) \Rightarrow (x = z))$,

(*d*) if $u = v$ and $R\{u\}$ and $R\{v\}$ are obtained by replacing the letter x in $R\{x\}$ by u and v respectively, then

$$R\{u\} \Leftrightarrow R\{v\}.$$

Axiom systems

An abstract deductive science is constructed by selecting an **axiom system,** i.e. certain undefined terms (e.g. 'set' in the Zermelo–Fraenkel Axiom system (p. 213)) and a number of axioms (relations of the specific 'once-and-for-all' type (p. 2)) containing them. From these axioms and using the rules of reasoning derived from the logical axioms, one obtains further relations which are *true* in the sense described on p. 3 and which are known as **theorems** or **lemmas** (the term used depends upon the extent to which mathematicians will wish to refer to them). The question of the 'truth' of the axioms is irrelevant within the framework of the axiom system. If, however, we can assign meanings to the undefined terms in such a way that the axioms can be judged 'true', then we say that we have a **model** or an **interpretation** of the abstract axiom system and the system is said to be **satisfiable.**

The undefined terms appearing in an axiom system will, in general, be of two kinds, namely, the *universal terms* such as 'set' and '$\in$' (p. 8), and the *technical terms* peculiar to that system (sometimes known as *primitive terms*) such as 'point' and 'line'.

Since our aim is to model rational thought, the most fundamental question to be asked about an axiom system is 'Does the system imply any contradictory theorems?' (i.e. is there a relation R such that both R and $\sim R$ can be deduced from the axioms?).

It can be shown that if a relation R exists such that both R and $\sim R$ can be deduced from the axioms, then any relation Q can be so deduced.

An axiom system Σ which does not imply contradictory statements is said to be **consistent** (see p. 23 for an example).

The usefulness of this definition is limited by our ability to recognise contradictions. It is not possible, for example, to show that the axioms of mathematics are consistent, yet all mathematicians proceed in the belief that they are. This belief is made explicit, for example, in the use of proofs based on **reductio ad absurdum**, i.e. proofs in which we infer the falsity of a relation P by adjoining it to our axiom system and then showing that in the new system there is some proposition Q such that both Q and $\sim Q$ are deducible.

Since there is, in general, no test for consistency we must be satisfied with what is termed **relative consistency**, obtained by showing that our axiom system Σ can be embedded in a second system Σ' (in whose consistency one is willing to believe). We take, therefore, as a 'working definition' of consistency: an axiom system is consistent if it is satisfiable.

If Σ is an axiom system and A is one of its axioms, then A is said to be **independent** in Σ, or an **independent axiom** of Σ, if it cannot be derived from other axioms in the system. This will be the case if both Σ and the axiom system obtained from Σ by replacing A by its negation (denoted by $(\Sigma - A) + (\sim A)$) are satisfiable (see p. 23 for an example).

An axiom system Σ is said to be **independent** if all the axioms of Σ are independent in Σ.

An axiom system Σ is said to be logically **complete** if it is impossible to add a new independent axiom to Σ which is consistent with Σ and which contains no new undefined terms, i.e. if there is no Σ-*statement* A such that A is an independent axiom of the system $\Sigma + A$.

An axiom system Σ is said to be **categorical** if every two models of Σ are isomorphic with respect to Σ, i.e. if there exists a one-to-one correspondence between the elements and relations of one model and those of the other such that whenever a given relation holds between two elements of one model, the corresponding relation holds between the corresponding two elements of the other.

Note. Categoricity implies logical completeness and so provides a means whereby the latter can be tested.

Example. The axioms for a group (see (*a*), (*b*), (*c*), p. 25) are not *categorical*, since it is not true that all groups are isomorphic, indeed it is possible to add further *independent* axioms to the system, for example, (*d*) the operation $*$ is commutative. We note though that if the axioms on p. 25 were changed so as to read '...consisting of a set G containing 7 elements and a binary operation $*$...', then the axiom system would be *categorical* since all models of the system, i.e., 'groups of order 7', are isomorphic.

2 Sets and functions

Sets

A set is a totality of certain definite, distinguishable objects of our intuition or thought – called the *elements* of the set.

This classic definition of a **set** was given by Georg Cantor in 1874. Such attempts to give elementary definitions of a set are, however, doomed to failure, their being in the main based on the use of undefined synonyms, such as 'collection', and leading to logical inconsistencies (see *Russell paradox* p. 210). For this reason, mathematicians now regard the notion of a set as an undefined, primitive concept (see *Zermelo–Fraenkel Axioms* p. 213).

Along with the idea of a 'set' we introduce the further basic mathematical sign of **membership** ($\in$).

$a \in A$ is then read as 'a is an **element (member)** of the set A' or 'a **belongs to** A'.

We write $a \notin A$ to signify that a is *not* an element of A.

Two sets, A and B, are equal (written $A = B$) if and only if they have the same elements, i.e.

$$\forall x, x \in A \Leftrightarrow x \in B.$$

B is said to be a **subset** of A (or is **included** in A) when every element of B is an element of A, i.e.

$$\forall x, x \in B \Rightarrow x \in A.$$

We denote that B is a subset of A by writing $B \subset A$.

If $B \subset A$ and $B \neq A$, we say that B is a **proper** subset of A.

Note. Some authors denote inclusion by $\subseteq$ and reserve the use of the sign $\subset$ to cases where the subset is proper.

If X is a set, then we denote the subset, A, of those elements $x \in X$ which satisfy a particular relation $R\{x\}$ by

$$A = \{x | x \in X, R\{x\}\} \quad \text{or, more concisely,} \quad A = \{x \in X | R\{x\}\}.$$

(It is an axiom that A is a set, see Zermelo–Fraenkel Axiom V, p. 214.)

Alternative notations are

$$A = \{x : x \in X, R\{x\}\}, \quad A = \{x \in X : R\{x\}\}.$$

If B is a subset of A, then we define the **complement** of B in A to be

$$\{x \in A | x \notin B\}.$$

(It follows that the complement will again be a set.)

The complement of B in A is denoted by $A-B$, $A \backslash B$, $\complement_A B$ or, when there is no ambiguity concerning the set A, by B' or $B^{\sim}$.

Note. Many authors, particularly when writing on Boolean algebra and measure theory, write $A-B$ to denote $\{x \in A | x \notin B\}$ but do not imply thereby that $B \subset A$ (see the definition of *difference* below).

The complement in A of the set A itself is called the **empty** subset of A. This set has no elements and is independent of A, i.e. $A-A = B-B$. It is called the **empty (null, void) set** and is denoted by $\varnothing$.

$\varnothing$ is a subset of every set X.

The set containing the single element x is denoted by $\{x\}$. A set having exactly one element is called a **singleton.**

Similarly, the set with elements $a, b, c, \ldots, z$ is denoted by $\{a, b, c, \ldots, z\}$.

Given a set A we denote the set whose elements are the subsets of A by $\mathscr{P}(A)$ or 2^A (see Zermelo–Fraenkel Axiom IV, p. 213). $\mathscr{P}(A)$ is called the **power set** or **set of subsets** of A.

Let A and B be two subsets of some set X.

(*a*) The **intersection** of A and B, written $A \cap B$ or AB, is the set of all those elements which belong to both A and B, i.e.

$$A \cap B = \{x \in X | x \in A \text{ and } x \in B\}.$$

A and B are said to be **disjoint** if they have no elements in common, i.e., if

$$A \cap B = \varnothing.$$

(*b*) The **union** of A and B, written $A \cup B$ or $A+B$, is the set of all those elements which belong to A or to B (or to both), i.e.

$$A \cup B = \{x \in X | x \in A \text{ or } x \in B\}.$$

The definitions of *intersection* and *union* can be extended to cover the intersection or union of a *family* of sets.

If I is any set, then we say that $\{A_i\}_{i \in I}$ is a **family of sets indexed by** I, the **index set**, if, corresponding to each $i \in I$, there is an associated set which we denote by A_i.

The **intersection** of a non-empty family of sets $\mathscr{A} = \{A_i\}_{i \in I}$ (i.e. $I \neq \varnothing$) is defined to be the set comprising those elements x which satisfy $x \in A_i$ for all $i \in I$. This intersection is denoted by

$$\bigcap_{i \in I} A_i \text{ or } \bigcap_{A \in \mathscr{A}} A.$$

Similarly, the **union** of the family of sets $\mathscr{A} = \{A_i\}_{i \in I}$ is defined to be the set of all those elements which satisfy $x \in A_i$ for some $i \in I$. This union is denoted by

$$\bigcup_{i \in I} A_i \text{ or } \bigcup_{A \in \mathscr{A}} A .$$

Note. (i) The use here of Zermelo–Fraenkel Axiom III (p. 213).

(ii) The fact that the use of i is not significant; any other letter may be used in its place (see also p. 134). For this reason we refer to i as being a *dummy suffix*.

(*c*) The **difference** of A and B is the set

$$\{x \in X \mid x \in A, x \notin B\}.$$

If $B \subset A$, the difference is said to be **proper**. If $B \not\subset A$, then the difference is sometimes referred to as the **relative complement** of B in A. As noted above, the difference is often denoted by $A - B$.

(*d*) The **symmetric difference** of A and B, $A \triangle B$, is

$$A \triangle B = (A \cap \mathbf{C}_X B) \cup (\mathbf{C}_X A \cap B),$$

or, in terms of *relative complements*,

$$A \triangle B = (A - B) \cup (B - A).$$

Example. Let $A = \{3, 6, 9, 12\}$ and $B = \{2, 4, 6, 8, 10, 12\}$.

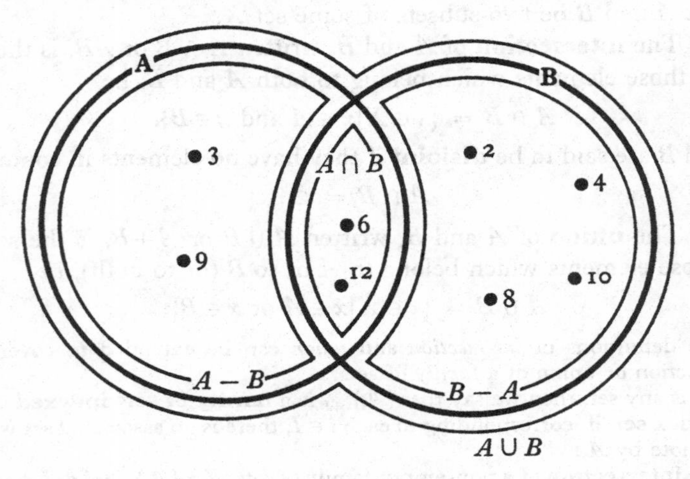

Then $\quad A \cap B = \{6, 12\}$ – the *intersection* of A and B,

$A \cup B = \{2, 3, 4, 6, 8, 9, 10, 12\}$ – the *union* of A and B,

$(A \cup B) - A = \{2, 4, 8, 10\}$ – the *complement* of A in $A \cup B$,

$$A-B = \{3, 9\} \text{ – the } \textit{relative complement} \text{ of } B \text{ in } A,$$

$$B-A = \{2, 4, 8, 10\} \text{ – the } \textit{relative complement} \text{ of } A \text{ in } B,$$

$$A \triangle B = \{2, 3, 4, 8, 9, 10\} \text{ – the } \textit{symmetric difference} \text{ of } A \text{ and } B.$$

The various sets can be readily distinguished in the diagrammatic representation. Diagrams of this type in which a set is represented by the region interior to a closed curve are called *Venn* (or *Euler*) *diagrams*.

Graphs and functions

Given two elements, a and b, of some set we define the **ordered pair** (a, b) by

$$(a, b) = \{\{a\}, \{a, b\}\}.$$

Note. (i) It follows from the definition that $(a, b) = (c, d)$ if and only if $a = c$ and $b = d$.

(ii) An alternative and more useful working definition is to take the ordered pair (a, b) to be some third object subject to the rule

$$(a, b) = (u, v) \Leftrightarrow a = u \quad \text{and} \quad b = v.$$

If $z = (x, y)$ is an ordered pair, then x and y are determined by z and we write

$$x = \text{pr}_1(z), \quad y = \text{pr}_2(z).$$

Here, pr denotes '*projection*' and derives from a geometrical interpretation of an ordered pair. An alternative notation, which is more frequently used in work in analysis (see, for example, p. 130) is

$$x = \pi^1(z), \quad y = \pi^2(z).$$

A set whose elements all consist of ordered pairs is called a **graph**. If G is a graph, then the idea of projection can be extended to give $\text{pr}_1(G) = X$, where

$$x \in X \Leftrightarrow \exists z \in G \text{ such that } x = \text{pr}_1(z).$$

The **Cartesian product**, $A \times B$, of two sets A and B is defined to be the set of all ordered pairs (a, b) of elements from A and B respectively, thus

$$z \in (A \times B) \Leftrightarrow \exists a \in A \text{ and } \exists b \in B \text{ such that } z = (a, b),$$

or

$$A \times B = \{(x, y) | x \in A, y \in B\}.$$

A Cartesian product may be projected to give

$$\text{pr}_1(A \times B) = A, \quad \text{pr}_2(A \times B) = B.$$

The idea of ordered pairs can be extended to give 'ordered triples', 'ordered quadruples', etc. by defining the ordered triple (a, b, c) to be $(a, (b, c))$.

We write $A \times B \times C$ to denote the set of ordered triples (a, b, c) such that $a \in A$, $b \in B$, $c \in C$. In this way we can extend the definition of Cartesian product to products of several factors. (In doing so we make conventions about the identification of such sets as $A \times (B \times C)$ and $(A \times B) \times C$.)

If A is a set, we write

$$A \times A = A^2, \quad A \times A \times A = A^3, \quad \text{etc.}$$

A **function** is defined to be an ordered triple

$$f = (G, A, B)$$

where G, A, B are sets which satisfy:

(i) $G \subset A \times B$,

(ii) for each $a \in A$, there is exactly one $b \in B$ such that $(a, b) \in G$.

We then call:

G the **graph of the function,**
A the **domain** of the function,
B the **codomain (range, target)** of the function.

Note. (i) Some authors use **range** to denote the subset of B defined by $\mathrm{pr}_2(G)$.

(ii) This definition of a function has arisen as a result of a series of generalisations and refinements. The word function was originally used only with regard to particular examples such as powers, trigonometric functions, etc. Euler, in the middle of the eighteenth century, used the word in two ways. The first referred to expressions in x which were made up of powers, logarithms, trigonometric functions and such like. His second type of function, $y(x)$, was defined for him whenever a curve was arbitrarily drawn in the x–y plane. A more general concept of function was formulated by Dirichlet whose definition ran 'If in any way a definite value of y is determined corresponding to each value of x in a given interval, then y is called a function of x'. Only in the last century has the notion of function been extended to cover domains which were not subsets of the real or complex numbers. A function then came to be understood as a mathematical object, f, associated with two sets A and B, which assigns to each element $a \in A$ a unique element $f(a) \in B$, e.g. the function which assigns to each plane triangle a number known as the area of the triangle. As the notion of function has been refined there has been the need to create words to describe particular types of function, see, for example, analytic functions (p. 154), and even to discard descriptions which did service for many years, such as 'many-valued functions'.

It follows from the definition that two functions f and g will be **equal** (in the sense of equality of sets and triples) if and only if they have the same domain and codomain and $f(a) = g(a)$ for all $a \in A$.

If A, B are two sets, then a **mapping** of A **into** B is a function, f, with A as domain and B as codomain. We say 'f **maps** A **into** B'.

'Mapping' and 'function' can, therefore, be regarded as synonyms. Topologists do, however, use the term 'mapping' in a more restricted sense to mean 'continuous mapping' (see p. 103).

If f maps A into B, then we write

$$f: A \to B \quad \text{or} \quad A \xrightarrow{f} B.$$

If $a \in A$ and $(a, b) \in G$, i.e. the mapping f assigns to a the element $b \in B$, then we write

$$f(a) = b \quad \text{or} \quad f: a \mapsto b.$$

b is called the **value** of the function f corresponding to the **argument** a.

Note the use of the barred arrow to go from argument to value compared with that of the straight arrow to go from domain to codomain. This distinction is not observed by all authors.

Given $f = (G, A, B)$ and $X \subset A$, then we denote the subset of B which consists of values of f corresponding to elements of X by $f(X)$, i.e.

$$f(X) = \{b \mid b = f(x) \text{ for some } x \in X\}.$$

In particular, $f(A)$ is sometimes denoted by $\text{Im}(A)$ (or $\text{Im}(f)$) and described as the **image** (or, as mentioned above, the **range**) of f. A function whose image is a singleton (p. 9), is known as a **constant function.**

Given a set A and a set B it is possible to define many functions which map A into B. The **set of all possible mappings of A into B** is denoted by B^A or $\mathscr{F}_B(A)$ or $\mathscr{F}(A, B)$.

The fact that a subset X of A can be characterised by that mapping of A onto $\{0, 1\}$ which maps all elements of X onto 1 and all elements of $A - X$ onto 0, accounts for the similarity between the notation B^A and the alternative notation, 2^A, for the power set of A (p. 9).

A function $g = (G', A', B')$ is said to be a **restriction** to A' of $f = (G, A, B)$ if

$$A' \subset A, \quad B' \subset B \quad \text{and} \quad g(a) = f(a) \quad \text{for all} \quad a \in A'.$$

Similarly, f is an **extension** of g to A if

$$A' \subset A, \quad B' \subset B \quad \text{and} \quad g(a) = f(a) \quad \text{for all} \quad a \in A'.$$

Example. We can define a *function* f with *domain* $\mathbb{N}$ (p. 20) and *codomain* $\mathbb{N}$, i.e. $f: \mathbb{N} \to \mathbb{N}$, by $f(n) = n^2$, i.e. $f: n \mapsto n^2$. n^2 is the *value* of f corresponding to the *argument* n. The *image* of f is a proper subset of the codomain, namely the set $\{0, 1, 4, 9, 16, \ldots\}$. The function g with domain $\{2, 4, 6, 8, \ldots\}$ and codomain $\mathbb{N}$ defined by $g: n \mapsto n^2$ is a *restriction* of f, and $h: \mathbb{R} \to \mathbb{R}$ (p. 65) defined by $h: x \mapsto x^2$ is an *extension* of f.

Some special types of function

Let f be a mapping of A into B. We say that f is:

surjective (onto) when $f(A) = B$;
injective (one–one) when $\forall a, a' \in A, f(a) = f(a') \Rightarrow a = a'$;
bijective (one–one onto) when it is both injective and surjective.

f is then known respectively as a **surjection**, an **injection** or a **bijection.**

Example. The function $f: \mathbb{N} \to \mathbb{N}$ defined in the example above is *injective* but not *surjective*. $f: \mathbb{N} \to S$ defined by $n \mapsto n^2$ where $S = \{0, 1, 4, 9, \ldots\}$ is *bijective* since it is both *injective* and *surjective*.

A bijection of a set A onto itself is called a **permutation** of A.

For any set A, the **identity function,** $1_A(id_A, j_A)$: $A \to A$, is the function $a \mapsto a$ which maps every element onto itself.

If $B \subset A$, then the function $i: B \to A$ which maps every element of B onto itself (as an element of A) is called the **insertion** of B into A.

Composition of functions

Let $f: A \to B$, $g: B \to C$ be mappings of A into B and B into C respectively. We construct a third mapping $h: A \to C$ which is called the **composite (composition)** of f and g and is denoted by $g \circ f$ (or gf), by setting

$$h(a) = g(f(a)) \quad \text{for all} \quad a \in A.$$

Note. (i) In the past the composite of two functions was often referred to as a *function of a function*.

(ii) The way in which authors indicate the composite map 'first f and then g' varies. Plainly it would be preferable to write it $f \circ g$ so that one would 'read' the functions in their natural order. The usage given here, however, has certain advantages in analysis, not least that the writing of composites in the form $x \mapsto \sin 3x$ and $x \mapsto \log \sin x$ can be found in almost every text. In algebra this usage is not so firmly established and in the case of some topics, e.g. permutations, the $f \circ g$ form is preferred. Indeed, to facilitate this usage, the notation xf is sometimes employed instead of $f(x)$. 'First f and then g' is then written xfg which leads naturally to the use of $f \circ g$ to denote the composite function. (See also p. 95, note (ii).)

(iii) Care must be taken when the gf notation is being used to distinguish between the *composite* and the *product* of two functions, i.e. given $f: x \mapsto 3x$ and $g: x \mapsto \sin x$, one must distinguish carefully between the *composite* $x \mapsto \sin 3x$ and the *product* $x \mapsto 3x \sin x$ (see p. 116). Such confusion occurs most frequently in analysis, and for that reason analysts prefer to denote the composite mapping by $g \circ f$ – algebraists tend to use the more compact notation gf.

Composition of functions can be represented diagrammatically.

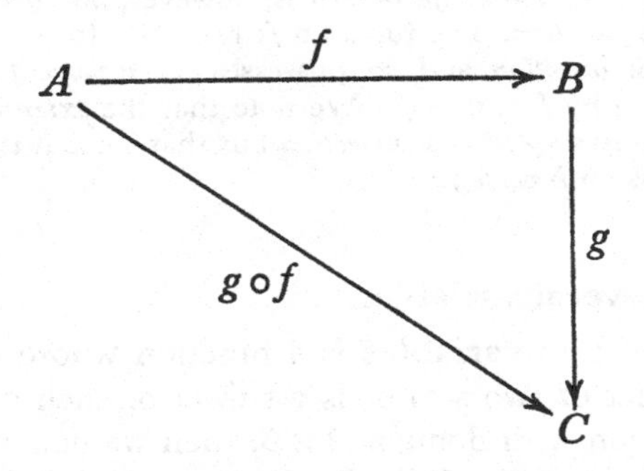

In the diagram we see that going to C from A in one step by the composite $g \circ f$ is equivalent to going in two steps via B: we say that the triangular diagram **commutes.**

Inverse images and inverse functions

Given $f: A \to B$, we define the **inverse image (counter image)** of $X \subset B$ to be the subset of A:

$$\{a \mid a \in A \quad \text{and} \quad f(a) \in X\}.$$

This subset is often denoted by $f^{-1}(X)$.

Given $f: A \to B$ and $g: B \to A$, we say that g is a **left inverse** of f and f a **right inverse** of g if $g \circ f: A \to A$ is the identity 1_A. If, moreover, $f \circ g: B \to B$ is the identity 1_B, then we say that f is a **two-sided inverse** of g (g will then be a two-sided inverse of f).

It can be shown that f will have a two-sided inverse if and only if it is a bijection, in which case the two-sided inverse is unique and bijective and can therefore be referred to without ambiguity as *the* **inverse function** of f, denoted by f^{-1}.

Note. It will be observed that f^{-1} has been used to denote two functions:

(*a*) the *inverse function* of $f: A \to B$, which when f is a bijection is a function with domain B and codomain A; and

(*b*) the *inverse-image function* which has domain $\mathscr{P}(B)$ and codomain $\mathscr{P}(A)$.

In order to prevent possible confusion some authors use an alternative notation, e.g. f^p, for this latter function.

Example. As on p. 14 we consider the function $f: \mathbb{N} \to \mathbb{N}$ defined by $f(n) = n^2$. The *inverse image* of, say, $\{2\} \subset \mathbb{N}$ will be $f^{-1}(\{2\}) = \varnothing$ and that of $\{4\}$ will be $f^{-1}(\{4\}) = \{2\}$. If we define $f^*: \mathbb{N} \to \mathbb{N}$ by $n \mapsto \sqrt{n}$ if $\sqrt{n} \in \mathbb{N}$ and $n \mapsto 1$ otherwise then $f^* \circ f: \mathbb{N} \to \mathbb{N}$ is the identity $1_{\mathbb{N}}$, i.e. f^* is a *left inverse* of f. However, $f \circ f^*: \mathbb{N} \to \mathbb{N}$ is *not* the identity since if $\sqrt{n} \notin \mathbb{N}$ then $f(f^*(n)) = 1$. Since f is *not* a *bijection* (it is, however, an *injection*) it does *not* possess an *inverse function*. The function $f: \mathbb{N} \to S = \{0, 1, 4, 9, \ldots\}$ defined by $f: n \mapsto n^2$ is a *bijection* and so possesses an *inverse function*, namely $f^{-1}: S \to \mathbb{N}$ defined by $f^{-1}: n \mapsto \sqrt{n}$. We note that the *extension*, $\bar{f}$, of f to $\mathbb{Z}$, $\bar{f}: \mathbb{Z} \to S$, defined by $x \mapsto x^2$ is a *surjection* but that since it is *not* an *injection* it does *not* possess an *inverse function*.

Functions of several variables

A **function of two variables** is a function whose domain is the Cartesian product of two sets or is a subset of such a product.

If f is a function with domain $A \times B$, then we denote the value of f at the point (a, b) of $A \times B$ by $f(a, b)$.

The notion can be generalised to define functions for which the domain and codomain are products, for example,

$$f: A \times B \to X_1 \times X_2 \times X_3$$

where A, B, X_1, X_2, X_3 are arbitrary sets.

Using previously defined notation (p. 11) we have

$$\mathrm{pr}_i(X_1 \times X_2 \times X_3) = X_i \quad (i = 1, 2, 3)$$

and can therefore define $f_i (= \mathrm{pr}_i \circ f): A \times B \to X_i$.

To determine f it is necessary and sufficient to know the three functions f_1, f_2, f_3 and we write

$$f = (f_1, f_2, f_3).$$

Example. Let $A = \{r \in \mathbb{R} | r > 0\}$, $B = \{\theta \in \mathbb{R} | 0 \leqslant \theta < 2\pi\}$.

Then

$$f: A \times B \to \mathbb{R} \times \mathbb{R},$$

defined by $f: (r, \theta) \mapsto (r\cos\theta, r\sin\theta)$, is a *function of two variables* (which, incidentally, tells us how to convert from *polar coordinates* to *Cartesian coordinates* (p. 69)).

$$\mathrm{pr}_1 \circ f = f_1 \quad \text{is the function} \quad (r, \theta) \mapsto r\cos\theta,$$
$$\mathrm{pr}_2 \circ f = f_2 \quad \text{is the function} \quad (r, \theta) \mapsto r\sin\theta.$$

Operations

A function $f: X \times X \to X$ is called a **binary operation on** the set X.

It is usual in the case of functions which are binary operations to denote the *value* of the function by some notation other than $f(x, y)$. Thus, in the case of the *binary operation* of addition defined on the real numbers $\mathbb{R}$, i.e., the function $+: \mathbb{R} \times \mathbb{R} \to \mathbb{R}$, the value of the function is denoted by $x+y$ rather than $+(x, y)$.

More generally we define an **n-ary operation on** the set X to be a function

$$f: X^n \to X \quad (n > 0).$$

A *nullary* operation ($f: \{a\} \to X$) consists of selecting an element of X.

A binary operation $*: X \times X \to X$, denoted by $(x, y) \mapsto x * y$, is said to be:

(i) **commutative** when $x * y = y * x$ for all $x, y \in X$;

(ii) **associative** when $(x * y) * z = x * (y * z)$ for all $x, y, z \in X$.

An element e is said to be a **left identity (unit)** for $*$ if $e * x = x$ for all $x \in X$. e is called a **right identity** if $x * e = x$ for all $x \in X$. e is called an **identity (unit) element** for $*$ if it is both a right and a left identity.

Note. The use of the term 'unit' can lead to confusion since, see p. 30, 'unit' has another well-defined meaning.

A second operation $\dagger$ is said to be **distributive over** $*$ when

$$x \dagger (y * z) = (x \dagger y) * (x \dagger z) \quad \text{for all} \quad x, y, z \in X.$$

Given a binary operation $*$ on a set X, we say that a subset Y of X is **closed under** $*$ if and only if

$$\forall x, y \in Y, \quad x * y \in Y.$$

Examples. The function $\mathscr{P}(A) \times \mathscr{P}(A) \to \mathscr{P}(A)$ (p. 9) defined by

$$(B, C) \mapsto B \cap C$$

is a *binary operation* on $\mathscr{P}(A)$. As mentioned above, it is usual to denote the *value* of this function by $B \cap C$ rather than $\cap(B, C)$. We note that $\cap$ is both *commutative* and *associative* and that A is an *identity element* for $\cap$ since $A \cap B = B \cap A = B$ for all $B \in \mathscr{P}(A)$. Moreover, the *binary operation* $\cup$ is *distributive* over $\cap$.

The operation of multiplication defined on the real numbers is a *binary operation* and the subset $\{x \in \mathbb{R} | 0 < x < 1\}$ is *closed under* multiplication.

See also: Axiom of Choice (p. 201); De Morgan's Laws (p. 202); Galileo's paradox (p. 203); Russell's paradox (p. 210); Schröder–Bernstein Theorem (p. 210).

3 Equivalence relations and quotient sets

Binary relations

Given any two sets X and Y, a subset $R \subset X \times Y$ (i.e. a graph) is called a **binary relation**. If $(x, y) \in R$, then we write xRy, Rxy or $R\{x, y\}$.

In a similar manner, by considering subsets of Cartesian products of n factors, we can define ***n*-ary relations**.

It follows from this definition that the graph of every function of X into Y is a binary relation. It is not true, however, that every such relation is the graph of a function since it is possible that for some $z \neq y$ one has xRy and xRz.

Let R be a binary relation defined on a set X to itself (i.e. R is a subset of $X \times X$). Then we say R is:

(i) **reflexive** when xRx for all $x \in X$;

(ii) **symmetric** when xRy implies yRx for all $x, y \in X$;

(iii) **transitive** when xRy and yRz together imply xRz for all $x, y, z \in X$.

A binary relation on a set X (to itself) which is reflexive, symmetric and transitive is said to be an **equivalence relation** on X.

R is said to be:

(iv) **antisymmetric** when for all $x, y \in X$, xRy and yRx together imply $x = y$.

A binary relation on a set X which is reflexive, transitive and antisymmetric is called a **partial order** of X. A set with a partial order is known as a **poset** (see p. 75).

Examples. Let R be the *binary relation* defined on $\mathbb{Z}$ (p. 23) by nRm if and only if $3|m-n$ (3 is a divisor of $m-n$). Then R is an *equivalence relation* since it satisfies conditions (i), (ii) and (iii). R is *not* a function since, for example, $2R8$ and $2R11$, i.e. $(2, 8)$ and $(2, 11)$ belong to the *graph* of the relation (this contradicts condition (ii) in the definition of a function on p. 12). R is *not* antisymmetric since we have $2R8$ and $8R2$ yet $2 \neq 8$. The relation $\subset$ defined on the set $\mathscr{P}(A)$ provides an example of a *partial order*. Thus $(\mathscr{P}(A), \subset)$ is an example of a *poset*.

Quotient sets

Suppose we are given a function $f: A \to B$. We can then define an equivalence relation on A, called the **equivalence kernel** of f and denoted by E_f, by

$$xE_f y \Leftrightarrow f(x) = f(y).$$

In the reverse direction, given an equivalence relation E on a set A, we define the **equivalence class** under E of any element $a \in A$ to be

$$p_E(a) = \{x \in A \mid xEa\}.$$

An alternative notation for xEa is $x \equiv a \pmod{E}$. If this notation is followed it is usual to talk of the equivalence class under E of a as the **class of a modulo E.**

Any subset of A which is the equivalence class under E of any element is called an **equivalence class** for E (E-class). The subset of $\mathscr{P}(A)$ consisting of all the equivalence classes for E is called the **quotient set of A by E** and is denoted by A/E.

Since $x \mapsto p_E(x)$ assigns to each $x \in A$ an E-class, it is a function $p_E: A \to A/E$. It is known as the **canonical map (projection)** of A onto its quotient by E.

The quotient set is a *partition* of A when we define a **partition, π, of a set X** to be a subset of $\mathscr{P}(X)$ such that:

(i) for each $x \in X$, x is an element of exactly one $S \in \pi$;

(ii) $\varnothing \notin \pi$.

Example. Let R be the equivalence relation defined in the example on p. 18. The *equivalence classes* under R are the sets

$$A = \{\ldots, -6, -3, 0, 3, 6, \ldots\},$$

$$B = \{\ldots, -5, -2, 1, 4, 7, \ldots\},$$

$$C = \{\ldots, -4, -1, 2, 5, 8, \ldots\}.$$

If $x \in B$, say, then we write $x \equiv 1(R)$, $x \equiv 4(R)$, ... or, in this particular example, $x \equiv 1(3)$, $x \equiv 4(3)$, ...

The *quotient set* $\mathbb{Z}/R$ is the set $\{A, B, C\}$ and is a *partition* of $\mathbb{Z}$. The *canonical map* is the function $p_R: \mathbb{Z} \to \{A, B, C\}$ which maps every element onto the equivalence class to which it belongs, thus $p_R(1) = p_R(7) = B$.

4 Number systems I

There are two basic approaches to the definition of the set of **natural numbers,** $\mathbb{N} = \{0, 1, 2, 3, \ldots\}$, and of the arithmetical operations defined upon it: one starting point is the system of axioms due to Peano, the other a construction based upon set theory and the Zermelo–Fraenkel Axioms in which system the Peano Axioms will have the status of theorems.

Note. Historically, 0 was not regarded as a natural number and, indeed, in Peano's original development $\mathbb{N}$ was obtained as the set $\{1, 2, 3, \ldots\}$. A set-theoretic argument, however, leads more naturally to the inclusion of 0 in the set of natural numbers. For the sake of consistency, therefore, in what follows Peano's Axioms have been slightly modified so that they will yield the set $\mathbb{N} = \{0, 1, 2, 3, \ldots\}$.

Peano's Axioms

The system of natural numbers, $\mathbb{N}$, is a set $\mathbb{N}$ with a function $\sigma : \mathbb{N} \to \mathbb{N}$ (the **successor function**) and a distinguished element 0 such that:

(i) 0 is a natural number ($0 \in \mathbb{N}$);

(ii) σ is an injection ($\sigma(n) = \sigma(m) \Rightarrow n = m$);

(iii) for every $n \in \mathbb{N}$, $\sigma(n) \neq 0$ (there is no natural number having 0 as a successor);

(iv) any subset $U \subset \mathbb{N}$ with the two properties

$$(a)\ 0 \in U, \quad (b)\ \text{for all } n \in \mathbb{N},\ n \in U \Rightarrow \sigma(n) \in U,$$

must be the whole set $\mathbb{N}$. Postulate (iv) is called the **principle of mathematical induction.**

Addition and **multiplication** are defined on $\mathbb{N}$ by

$$m+n = \sigma^n(m), \quad m.n = (\sigma^m)^n(0) \quad (\text{all } m, n \in \mathbb{N}).$$

Here σ^n is the composite of n factors σ, i.e. it is defined by

$$\sigma^0 = 1_{\mathbb{N}}, \quad \sigma^{\sigma(n)} = \sigma \circ \sigma^n \quad (\text{all } n \in \mathbb{N}).$$

($1_{\mathbb{N}}$ denotes the identity function on $\mathbb{N}$, see p. 14.)

Inequalities are defined on $\mathbb{N}$ by

m is **smaller** (or **less**) than n ($m < n$) if and only if there exists $x \neq 0$ in $\mathbb{N}$ such that $n = m+x$.

We write

$$m \leqslant n \quad \text{for} \quad m < n \quad \text{or} \quad m = n,$$

$$m > n \quad \text{for} \quad n < m \quad (m \text{ is } \textbf{greater} \text{ than } n),$$

$$m \geqslant n \quad \text{for} \quad n \leqslant m.$$

The following results can then be obtained (by proofs which are far from easy):

(*a*) the **trichotomy** law: given $n, m \in \mathbb{N}$, then exactly one of

$$m < n, \quad m = n, \quad m > n$$

holds;

(*b*) addition and multiplication are *commutative* (p. 17) and *associative* (p. 17) operations, and multiplication is *distributive* (p. 17) over addition;

(*c*) addition and multiplication of elements of $\mathbb{N}$ are **isotonic**, i.e. for all $k, m, n \in \mathbb{N}$,

$$m < n \quad \text{and} \quad k \in \mathbb{N} \Rightarrow m+k < n+k,$$

$$m < n \quad \text{and} \quad k \in \mathbb{N} - \{0\} \Rightarrow mk < nk;$$

(*d*) $\mathbb{N}$ is *well-ordered* (p. 80), i.e. given $V \in \mathscr{P}(\mathbb{N})$ and $V \neq \varnothing$, then there is an element $f \in V$ such that $x \in V$ implies $x \geqslant f$ (f is the 'first element' of V).

In general, an **ordered set**, S, is a set together with a binary relation, $<$ say, which is transitive and satisfies the trichotomy law. Some authors use the term *ordered set* for what we have described as a *poset* (p. 18), in which case an ordered set for which the trichotomy law holds is called a **totally ordered set.**

A set-theoretic approach

We say that a set X is **equipotent** to a set Y, written $XEqY$ (or $Eq(X, Y)$), if there exists a bijection of X onto Y.

Given a set X we attach to it a new mathematical object, the **cardinal (number)** of X, written $\#(X)$ (Card (X)), defined in such a way that

$$XEqY \Leftrightarrow \#(X) = \#(Y).$$

Note. We avoid the obvious definition of $\#(X)$ as the equivalence class of X under Eq, since this, because of our definition of an equivalence relation, would imply that Eq was defined upon the paradoxical 'set of all sets'.

x is called a **cardinal number** if there exists some set X for which $x = \#(X)$.

We denote the cardinal of the empty set $\varnothing$ by the symbol 0, i.e. $0 = \#(\varnothing)$.

We denote the cardinal of the set $\{\varnothing\}$ by 1,
that of the set $\{\varnothing, \{\varnothing\}\}$ by 2,
that of the set $\{\varnothing, \{\varnothing\}, \{\{\varnothing\}\}\}$ by 3, etc.

Note. (i) $\#(X) = 2$ means that X is a set with two elements, i.e. $\exists x, y \in X$ such that $x \neq y$ and $z \in X$ implies either $z = x$ or $z = y$. (In this case $X Eq\{1, 2\}$ and this is what we mean by 'has two elements'.)
(ii) The use made of the Zermelo–Fraenkel Axiom (VII) (p. 214).

Given cardinal numbers x and y, we write

$$x \leqslant y$$

if there exist sets X and Y such that

$$x = \#(X), \quad y = \#(Y)$$

and X is equipotent to a subset of Y.

By the Schröder–Bernstein Theorem (p. 210) we then have: for any cardinal numbers x, y,
(i) either $x \leqslant y$ or $y \leqslant x$;
(ii) $x \leqslant y$ and $y \leqslant x$ together imply $x = y$.

Addition, **multiplication** and **exponentiation** of cardinal numbers are defined as follows:

Let x and y be two cardinal numbers and X and Y sets such that $x = \#(X)$, $y = \#(Y)$ and $X \cap Y = \varnothing$ (X and Y can always be defined so that this last condition is satisfied). Then we define:
(i) $x+y = \#(X \cup Y)$;
(ii) $xy = \#(X \times Y)$;
(iii) $x^y = \#(X^Y)$.

It must, of course, be verified that our definitions do not depend upon the choice of X and Y. Similar considerations will apply to many of the definitions given in this book. It will, however, be assumed in future that the reader realises that it is essential to check the consistency of such definitions.

It can be shown that

$$\#(\mathscr{P}(X)) = 2^{\#(X)},$$

and, moreover, that $x < 2^x$.

It follows, therefore, that there is no greatest cardinal number.

A set is said to be **finite** if the only subset of X which is equipotent to X is X itself. A set which is not finite is said to be **infinite**.

A cardinal number is **finite** if it is the cardinal of a finite set and is **infinite** otherwise.

A finite cardinal is called a **natural number**, and an infinite cardinal a **transfinite number.**

It can be shown that a set X is *finite* if and only if

$$\#(X) \neq \#(X)+1.$$

It follows that a *finite cardinal* (*natural*) *number* x is one for which

$$x \neq x+1.$$

It can now be shown that there exists a unique 'set of natural numbers' which is denoted by $\mathbb{N}$.

$\mathbb{N}$ is an infinite set and its cardinal is denoted by $\aleph_0$ (**aleph nought**).

A set A is said to be **countable** if it is equipotent to $\mathbb{N}$ or to a subset of $\mathbb{N}$. If it is equipotent to $\mathbb{N}$, then it is often described as **denumerable** or **enumerable.**

This distinction between *countable* and *enumerable* is not observed by all authors.

The cardinal $2^{\aleph_0}$, which we know to be strictly greater than $\aleph_0$, is denoted by c. It is called the **power of the continuum** and can be shown to be $\#(\mathbb{R})$ where $\mathbb{R}$ is the set of real numbers (p. 65).

The **continuum hypothesis** asserts that every infinite subset of $\mathbb{R}$ is equipotent to either $\mathbb{R}$ or $\mathbb{N}$. It was shown by Kurt Gödel (1906–) in 1940 that this hypothesis is *consistent* (p. 6) with the axioms of mathematics and in 1963 Paul Cohen (1934–) showed that it is *independent* (p. 6) of the other axioms, i.e. it, or its negation, can be added at will to the mathematician's axiom system.

The rational integers

The set of **rational integers** $\mathbb{Z} = \{\ldots, -3, -2, -1, 0, 1, 2, 3, \ldots\}$ can be constructed from $\mathbb{N}$ in the following way:

Define an equivalence relation E on $\mathbb{N}\times\mathbb{N}$ by

$$(x, y)E(x', y') \Leftrightarrow x+y' = x'+y$$

(we want the formula $x-y = x'-y'$ to hold once 'minus' has been defined). The set $\mathbb{Z}$ is then defined to be $(\mathbb{N}\times\mathbb{N})/E$. Given elements z and z' of $\mathbb{Z}$ such that

$$z = p_E(x, y), \quad z' = p_E(x', y'),$$

where p_E denotes the canonical mapping (p. 19) from $\mathbb{N}\times\mathbb{N}$ onto $\mathbb{Z}$, we define the sum and product of z and z' by

$$z+z' = p_E(x+x', y+y')$$
$$zz' = p_E(xx'+yy', xy'+x'y).$$

This definition is chosen because we want the formulae

$$(x-y)+(x'-y') = (x+x')-(y+y')$$

and $$(x-y)(x'-y') = (xx'+yy')-(xy'+x'y)$$ to hold.

It can be shown that with these two operations $\mathbb{Z}$ is a *commutative ring* (see p. 30). It is easily shown that the mapping $n \mapsto p_E(n, 0)$ is an injection of $\mathbb{N}$ into $\mathbb{Z}$ which preserves addition and multiplication. We can, therefore, identify $\mathbb{N}$ with a subset of $\mathbb{Z}$. Indeed, if we define the **negative** of z, written $-z$, to be the inverse element of z under addition (see p. 25), then it can be shown that either $z \in \mathbb{N}$ or $-z \in \mathbb{N}$. If $z \in \mathbb{N}-\{0\}$ we say that z is a **positive rational integer,** if $-z \in \mathbb{N}-\{0\}$ we say that z is a **negative rational integer.**

Note. (i) $a-b = a+(-b)$ if $a, b \in \mathbb{N}$.
(ii) The use of 'rational' in the description of $\mathbb{Z}$. This enables us clearly to distinguish between $\mathbb{Z}$ and the set of *algebraic* integers (see p. 72).

The rational numbers

The **set of rational numbers,** $\mathbb{Q}$, can be constructed in a similar manner to $\mathbb{Z}$, as follows.

Let E be the equivalence relation on $\mathbb{Z}\times(\mathbb{Z}-\{0\})$ defined by

$$(a, b)E(c, d) \Leftrightarrow ad = bc,$$

and define $\mathbb{Q}$ to be $$\mathbb{Z}\times(\mathbb{Z}-\{0\})/E.$$

Addition and **multiplication** are defined on $\mathbb{Q}$ in terms of the canonical mapping, p_E, by

$$p_E(a, b)+p_E(c, d) = p_E(ad+bc, bd),$$

$$p_E(a, b)\times p_E(c, d) = p_E(ac, bd).$$

It can be shown that with these operations $\mathbb{Q}$ is a *field* (p. 31) and that there is a natural injection which maps $\mathbb{Z}$ into $\mathbb{Q}$ and preserves the operations of multiplication and addition. We can, therefore, consider $\mathbb{Z}$ to be a subset of $\mathbb{Q}$.

Corresponding to each ordered pair (a, b) of $\mathbb{Z}\times(\mathbb{Z}-\{0\})$ is the fraction a/b with numerator a and non-zero denominator b (the need to make b non-zero accounts for the use of $\mathbb{Z}-\{0\}$, rather than $\mathbb{Z}$, in the definition). Two fractions are then equivalent if the corresponding ordered pairs are equivalent in the sense defined above, and a rational number is an equivalence class of fractions.

5 Groups I

A **group** is an ordered pair, $(G, *)$, consisting of a set G and a binary operation $*: G \times G \to G$ $((x, y) \mapsto x * y)$ satisfying

(*a*) $(x * y) * z = x * (y * z)$ for all $x, y, z \in G$ (*associativity*),

(*b*) there exists an element $e \in G$ such that $x * e = x = e * x$ for all $x \in G$,

(*c*) for each $x \in G$, there exists an element $x' \in G$ such that $x * x' = e = x' * x$.

The operation $*$ is frequently written as a product, i.e. $(x,y) \mapsto xy$, or as a sum, $(x, y) \mapsto x+y$. We then say that $(G, *)$ is respectively a **multiplicative** or an **additive** group.

The element e, which can be shown to be unique, is called the **identity** or **neutral element** of G. When G is multiplicative it is denoted by 1 and when G is additive by 0. Correspondingly, x', the **inverse element** of x (again unique) is denoted by x^{-1} and $-x$ respectively. (Cf. p. 15.)

Note that the definition of a group given above is wasteful in the sense that not all the conditions listed are independent (p. 6). Thus, for example, one can omit the condition $x = e * x$ from (*b*) and the condition $e = x' * x$ from (*c*), since these relations can be deduced from the remaining conditions.

Algebraic structures which do not satisfy all the group axioms occur frequently, and are given distinguishing names. For example:

An ordered pair $(S, *)$ consisting of a set S together with a binary operation $*: S \times S \to S$ which is associative, is known as a **semigroup**. A semigroup $(M, *)$ possessing an element e which satisfies group axiom (*b*) (i.e. which is an identity element for $*$) is called a **monoid**.

Note. (i) An ordered pair $(P, *)$ consisting of a set P together with a binary operation $*: P \times P \to P$ is sometimes known as a **groupoid**. It must be observed, however, that this word is also used in categorical algebra in an entirely different sense. (p. 95).

(ii) The hierarchy of structures can therefore be represented in tabular form as follows:

Set with binary operation} **groupoid** } **semigroup** } **monoid** } **group.**
Associativity
Identity element
Inverse elements

A group is said to be **commutative** or **Abelian** when the operation $*$ is commutative (see p. 17). It is usual to use the additive notation when considering Abelian groups.

Note. Some authors reserve the term *Abelian* for commutative groups written additively.

Examples. The set of integers $\mathbb{Z}$ is a *commutative group* under addition with *identity element* 0. The set of integers under multiplication is a *monoid* with identity element 1 but is *not* a group since it fails to satisfy axiom (*c*).

A group $(G, *)$ is said to be **finite** if the set G is finite, in which case $\#(G)$ (p. 21) is called the **order of the group.**

If g is any element of $(G, *)$, then we say that the **order (period) of the element** g is the smallest positive integer k which satisfies the relation

$$\underbrace{g * g * \ldots * g}_{k \text{ times}} = e.$$

If $*$ denotes $\times$, then the left-hand side of the relation is written g^k, if $*$ denotes $+$ it is written kg.

If no such integer exists, then we say that g has **infinite order** in $(G, *)$.

Given a prime p, a finite group G is called a **p-group** when the order of every element of G is some power of p.

Note. (i) It is customary in cases where no ambiguities can arise to denote a group by the same letter as the underlying set, i.e., by G, say, rather than $(G, *)$.

(ii) A finite group can be shown to be a p-group if and only if its order (i.e., the cardinal number of its underlying set) is some power of the prime p.

A group is said to be **periodic** when all its elements have finite order. A group in which no element other than the identity element has finite order is said to be **aperiodic** or **torsion-free.**

A **cyclic group** G is one containing an element x with the property that every other element of G can be written as a **power** of x, i.e. such that $\forall g \in G$, for some $n \in \mathbb{Z}$, $g = x^n$. We then say that G is the cyclic group **generated** by x.

Example. The *commutative group* $(\{0, 1, 2, 3, 4, 5\}; \oplus)$, where $a \oplus b$ is defined by $a \oplus b = c$ where $c \equiv a+b$ (6), known as the additive group of the integers modulo 6 and denoted by $(\mathbb{Z}_6, \oplus)$, is a group of *order* 6. Its *identity element* is 0 and the *orders* of the remaining elements 1, 2, 3, 4 and 5 are 6, 3, 2, 3 and 6 respectively. Moreover, $(\mathbb{Z}_6, +)$ is a *cyclic* group and is *generated* by 1 or -1 ($= 5$). We note that no element of the group $(\mathbb{Z}, +)$ other than 0 has finite order and so the group $(\mathbb{Z}, +)$ is *torsion-free*. Again $(\mathbb{Z}, +)$ is *cyclic* and is *generated* by 1 (or -1).

Given a set X, then the set $S(X)$ ($\mathfrak{S}(X)$) of all permutations of X (p. 14) with the binary operation $(f, g) \mapsto f \circ g$ is the **group of permutations of the set** X. In particular, the group of permutations of the set $\{1, 2, 3, \ldots, n\}$ is known as the **symmetric group,** S_n (Σ_n), of degree n. The group of all *even* permutations (p. 51) of n objects is known as the **alternating group** of degree n and is denoted by A_n ($\mathfrak{A}_n$).

Note. (i) The convention for denoting the composite of permutations is discussed in note (ii) on p. 14.

(ii) The *order* of S_n is $n(n-1) \ldots 2 . 1 = n!$ (p. 149), and that of A_n is $n!/2$.

If F is some figure (i.e., set of points) in the plane (p. 68), then the set of all permutations of F which preserve distance (p. 99) forms a group under the natural law of composition. This group is known as the **group of symmetries (symmetry group)** of F.

If, in particular, F is a regular polygon having n sides, then the group of symmetries (having order $2n$) is known as the **dihedral group** and is denoted by D_n(Δ_n or D_{2n}).

Example. The *symmetry group* of the square, D_4, is of order 8 with elements I, corresponding to the transformation which leaves all points unchanged; R_1, R_2 and R_3 corresponding to rotations about O through $\frac{1}{2}\pi$, π and $\frac{3}{2}\pi$ respectively; and M_1, M_2, M_3, M_4 corresponding to reflections in the axes AC, BD, XX' and YY'.

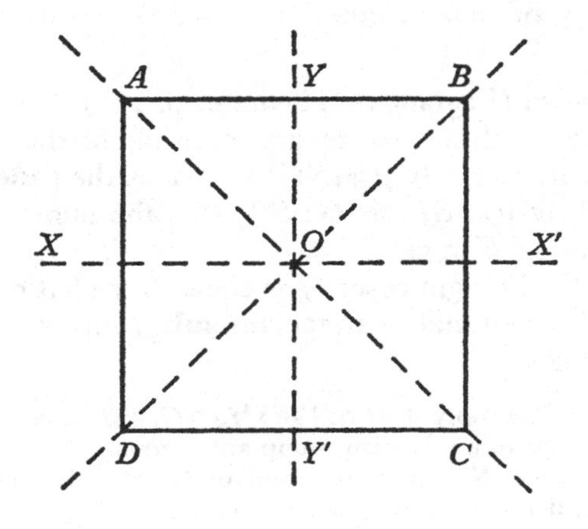

Subgroups

A non-empty subset S of a group G is said to be a **subgroup** of G if, for all $x, y \in S$: (*a*) $x * y \in S$, and (*b*) x^{-1} (or $-x$) $\in S$.

Thus S is a group, in which identity and inverse agree with those in G.

Subgroups of G other than G itself and $\{e\}$ are called **proper subgroups** of G.

If S is a subgroup of G, we say that G is an **extension** of S.

Note. A subgroup of S will be a subgroup of G.

Any subgroup T of a permutation group $S(X)$ is said to be a **transformation group** on the set X.

If S is a subgroup of G, then we can define an equivalence relation on the set G by

$$x\,E\,y \Leftrightarrow x^{-1}y \in S.$$

We can then construct equivalence classes in the natural way by defining

$$\mathrm{E}_x = \{y \in G | x\,E\,y\}.$$

E_x is, in fact, the set of all elements of G of the form xz where $z \in S$. We, accordingly, denote E_x by xS and refer to it as a **left coset** of S in G. Similarly, we can form the set of all elements of the form zx where $z \in S$ and, denoting this by Sx, refer to it as a **right coset** of S in G.

Note. (i) There is no universal convention regarding which set of cosets should be termed 'right' and which 'left'.

(ii) When G is an additive group the coset will be denoted by $S+x$ rather than Sx.

It can be shown (Lagrange's Theorem, p. 207) that the order of a finite group G is always an integral multiple of the order of any subgroup. The integer $\#(G)/\#(S)$ is known as the **index** of S in G and is denoted by $[G:S]$ (or $(G:S)$); it is the number of distinct left (right) cosets of S in G.

If for all $g \in G$, the right coset Sg is equal to the left coset gS, then we say that S is a **normal** or **invariant subgroup** of G and denote this by writing $S \lhd G$.

Note. (i) $S \lhd G$ if and only if $\forall s \in S$ and $\forall g \in G$, $gsg^{-1} \in S$.

(ii) All subgroups of an Abelian group are normal.

(iii) If $N \lhd G$ and S is *any* subgroup of G then the set of products $\{ns | n \in N, s \in S\}$, denoted by NS, is a subgroup of G.

Examples. $(\mathbb{Z}_6, \oplus)$, p. 26, has *proper subgroups*, $S = (\{0, 3\}, \oplus)$ and $T = (\{0, 2, 4\}, \oplus)$ both of which are *p-groups*. The *cosets* of S in $\mathbb{Z}_6$ are $\{0, 3\}$, $\{1, 4\}$, $\{2, 5\}$. Since the group $\mathbb{Z}_6$ is *commutative*, the *left* and *right cosets* will coincide, and S and T will be *normal subgroups*. The *index* of S in $\mathbb{Z}_6$ is 3 and $[\mathbb{Z}_6 : T] = 2$. D_4, p. 27, is non-commutative since

$$(M_3 =)\, R_1M_1 \neq M_1R_1 (= M_4).$$

$H = \{I, M_1\}$ is a *subgroup* of D_4 of index 4. The *left cosets* of H in D_4 are $IH = \{I, M_1\}$, $R_1H = \{R_1, M_3\}$, $R_2H = \{R_2, M_2\}$ and $R_3H = \{R_3, M_4\}$. The *right cosets* of H in D_4 are

$$HI = \{I, M_1\}, \quad HR_1 = \{R_1, M_4\}, \quad HR_2 = \{R_2, M_2\}$$

and

$$HR_3 = \{R_3, M_3\}.$$

Since $R_1H \neq HR_1$ the subgroup H is *not normal*. It can be checked that the subgroup $\{I, R_1, R_2, R_3\}$ is *normal* (as indeed is any subgroup S of index 2 in G).

See also: Cauchy's Theorem (p. 200); Cayley group table (p. 200); Hamiltonian group (p. 204); Lagrange's Theorem (p. 207); Sylow subgroup (p. 211); Sylow's Theorem (p. 211).

6 Rings and fields

A **ring** is an ordered triple $(R, +, .)$ consisting of a set R with two binary operations, + 'addition' and . 'multiplication', satisfying the following conditions:

(*a*) The pair $(R, +)$ is a commutative group;

(*b*) multiplication is associative and admits an identity (or unit) element, denoted by 1 (see note (i) below);

(*c*) multiplication is distributive (on both sides) over addition, i.e. $x.(y+z) = x.y+x.z$ and $(x+y).z = x.z+y.z$, for all $x, y, z \in R$.

Note. (i) It is becoming increasingly common not to include the second part of (*b*), i.e., the existence of a multiplicative identity element, amongst the **ring** axioms. One then distinguishes between **rings** and **rings with an identity (unit) element.** Similar differences also apply to the definitions of *subring* and *integral domain* given below.

(ii) It is customary to denote $x.y$ by xy.

If, moreover,

(*d*) multiplication is commutative,

then R is said to be a **commutative ring.**

Note. Commutative rings are often denoted by the letter K.

A **subring** of R is a subset S of R satisfying:

(*a*) S is a subgroup of the additive group R;

(*b*) $x \in S$ and $y \in S$ together imply $xy \in S$;

(*c*) $1 \in S$.

Thus a subring is a ring.

If an element $a \in R$ possesses an inverse element with respect to multiplication, i.e. if there exists a (unique) $a^{-1} \in R$ such that

$$aa^{-1} = a^{-1}a = 1,$$

then we say that a is an **invertible element** of R.

Note. A common alternative name for an invertible element is a *unit*. However, as mentioned earlier, the use of this latter nomenclature can lead to ambiguities.

The set of invertible elements of a ring R is denoted by R^*.

If every non-zero element of R is invertible, then R is said to be a **division ring (skew field)**.

A commutative division ring is called a **field**.

Note. If F is a field, then $F^* = F - \{0\}$. We insist that F^* should be non-empty, i.e., in a division ring or field we insist that $1 \neq 0$ – the underlying ring is *non-trivial.*

S is a **subfield** of F if S is a subring of F and $x \in S$, $x \neq 0$, together imply $x^{-1} \in S$.

A ring is **ordered** when there is a non-empty subset $P \subset R$, called the **set of positive elements** of R, satisfying:

(*a*) $a \in P$ and $b \in P$ together imply $a+b \in P$ and $ab \in P$;

(*b*) for each $a \in R$ exactly one of the following holds:

$$a \in P, \quad a = 0, \quad (-a) \in P.$$

By setting $a < b$ if $a, b \in R$ and $b - a \in P$, we retrieve our familiar ordering. Conversely, given an order relation $\geqslant$ on an integral domain (see below) we can define P to be $\{a \in R | a \geqslant 0, a \neq 0\}$.

A commutative ring is said to be an **integral domain** if $x \in K$, $y \in K$ and $xy = 0$ together imply $x = 0$ or $y = 0$.

Note. (i) If $xy = 0$, $x \neq 0$ and $y \neq 0$, then x and y are said to be **zero divisors** in K,

(ii) Not all authors insist that an integral domain shall be commutative.

(iii) The hierarchy of structures is indicated in the diagram – the 'weaker' structure, i.e., the one satisfying fewer axioms, being printed above the 'stronger' one.

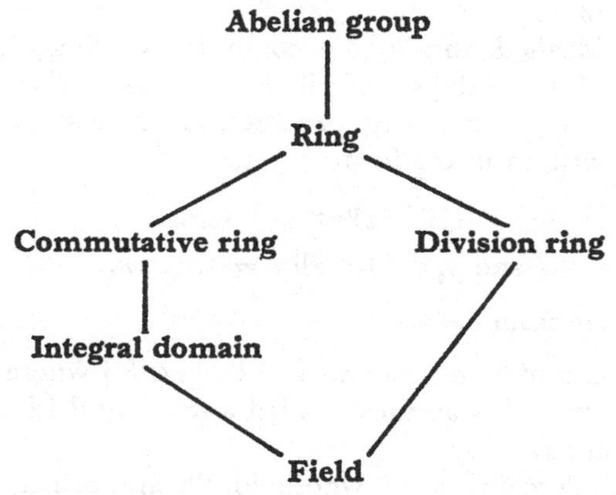

The **characteristic** of a ring is defined to be the additive order of its (multiplicative) identity element, i.e. R has finite characteristic m

if m is the least positive integer for which $m1 = 0$. It has characteristic 0 (or ∞, for usage differs) if no such multiple is zero.

Examples of *rings* are $(\mathbb{Z}, +, .)$, $(\mathbb{Q}, +, .)$ and $(\mathbb{Z}_6, \oplus, \otimes)$ (where $a \otimes b$ is defined to be $c \in \{0, 1, ..., 5\}$ such that $c \equiv a.b(6)$). $\mathbb{Z}$ is a *subring* of $\mathbb{Q}$, but it can be shown that $\mathbb{Z}$ and $\mathbb{Z}_6$ have no proper subrings. The only *invertible elements* of $\mathbb{Z}$ are 1 and -1 whereas all non-zero elements of $\mathbb{Q}$ are *invertible*. Since $\mathbb{Q}$ is a *commutative ring*, it follows that $\mathbb{Q}$ is a *field*. $\mathbb{Z}_n$ is a field only when n is a prime. $\mathbb{Q}$ is a *subfield* of $\mathbb{R}$. $\mathbb{Z}$, $\mathbb{Q}$ and $\mathbb{R}$ can all be ordered by $<$ used in its usual sense. $\mathbb{Z}$, $\mathbb{Q}$ and $\mathbb{R}$ are all *integral domains*. $\mathbb{Z}_6$ is not an *integral domain* since $3 \otimes 2 = 0$, i.e., 3 and 2 are *zero divisors*. $\mathbb{Z}_6$ has *characteristic* 6; $\mathbb{Z}$, $\mathbb{Q}$ and $\mathbb{R}$ have *characteristic* 0. $\mathbb{Z}_6$ cannot be *ordered*, for since either a^2 or $(-a)^2$ must be positive (by use of properties (*a*) and (*b*)) it follows that 1 is positive and so, since $0 = 1 \oplus 1 \oplus ... \oplus 1$ it follows (by (*a*)) that 0 is positive which contradicts (*b*). It can be shown similarly that any ordered ring must have characteristic 0.

A **(two-sided) ideal,** I, of a ring R is a non-empty subset of R satisfying:

(*a*) I is a subgroup of the additive group R,

(*b*) for all $x \in I$ and all $a, b \in R$ we have

$$\text{(i) } ax \in R \quad \text{and} \quad \text{(ii) } xb \in R.$$

I is a **left ideal** if it satisfies axioms (*a*) and (*b*)(i) and a **right ideal** if it satisfies (*a*) and (*b*)(ii). When R is commutative all ideals are two-sided.

An ideal, $I \neq K$, of a commutative ring K is said to be a **maximal ideal** of K if, whenever M is an ideal of K satisfying $I \subset M \subset K$, either $M = I$ or $M = K$.

Given two ideals I and J of a commutative ring K, we define their **sum,** $I+J$, to be the set of all elements of K of the form $x+y$ where $x \in I$ and $y \in J$, and their **product,** IJ, to be all elements of K which can be written in the form

$$x_1y_1 + x_2y_2 + ... + x_ny_n,$$

where $n \in \mathbb{N}$, $x_i \in I$ and $y_i \in J$ for all $i = 1, ..., n$.

$I+J$ and IJ are again ideals.

If K is a commutative ring and $I = \{xb \mid x \in K\}$ where b is a fixed element of K, then I is an ideal, called a **principal ideal.** b is then said to **generate** I.

An integral domain all of whose ideals are principal is called a **principal ideal domain** (or **principal ideal ring**).

If I is an ideal of a principal ideal domain D and is generated by the elements $x_1, \ldots, x_n$, i.e. I is the set of all elements of the form

$$u_1x_1 + u_2x_2 + \ldots + u_nx_n,$$

where $u_i \in D$, then I is generated by some single element d. We say that d is a **highest common factor (h.c.f.)** or **greatest common divisor (g.c.d.)** of $x_1, \ldots, x_n$.

$x_1, \ldots, x_n$ are said to be **mutually** or **relatively prime** if they have h.c.f. 1.

The intersection of the n ideals generated by $x_1, \ldots, x_n$ taken one at a time is also an ideal and so will be generated by a single element m. m is called a **least common multiple (l.c.m.)** of $x_1, \ldots, x_n$.

An element p of D is said to be **prime** or **irreducible** if it is not an invertible element of D and if $p = ab$, where $a, b \in D$, implies that a or b is invertible.

If $a, b, u \in D$ and $a = bu$ where u is invertible, then we say that a and b are **associates.** A prime is therefore divisible only by invertible elements and associates.

An integral domain D is a **unique factorisation domain** if

(*a*) $\forall a \in D - \{0\}$, a is either invertible or can be written as the product of a finite number of irreducible elements of D, and

(*b*) the decomposition in (*a*) is unique up to the ordering of the irreducible elements and substitution by associates.

Example. The sets

$$I_2 = \{\ldots, -4, -2, 0, 2, 4, \ldots\} \quad \text{and} \quad I_3 = \{\ldots, -6, -3, 0, 3, 6, \ldots\}$$

are *maximal ideals* of $\mathbb{Z}$.

$$I_{12} = \{\ldots, -24, -12, 0, 12, 24, \ldots\} \quad \text{and} \quad I_8 = \{\ldots, -16, -8, 0, 8, 16, \ldots\}$$

are *ideals* but *not* maximal ideals since; for example, $I_8 \subset I_2$ and $I_{12} \subset I_3$. $I_8 + I_{12}$ is the ideal $I_4 = \{\ldots, -8, -4, 0, 4, 8, \ldots\}$ and

$$I_8 I_{12} = I_{96} = \{\ldots, -192, -96, 0, 96, 192, \ldots\}.$$

$I_2, I_3, I_4, I_8, \ldots$ are all *principal ideals* and are generated respectively by 2, 3, 4, 8, ... $\mathbb{Z}$ is a *principal ideal domain.* The ideal generated by the elements 8 and 12 is the ideal $\{\ldots, -8, -4, 0, 4, 8 \ldots\} = I_4$. Hence, 4 is an *h.c.f.* of 8 and 12. (The other h.c.f. is -4. In general, any two different h.c.f.s of a given a and b are *associates.*) 8 and 11 together generate $\mathbb{Z}$ and so are *mutually prime* with *h.c.f.* 1. $I_8 \cap I_{12} = \{\ldots, -48, -24, 0, 24, 48, \ldots\} = I_{24}$ and so 24 is an *l.c.m.* of 8 and 12. (The other *l.c.m.* is the *associate* of 24, -24.) The *primes* of $\mathbb{Z}$ are $\pm 2, \pm 3, \pm 5, \pm 7, \ldots$; each prime being *associate* to a positive prime. $\mathbb{Z}$ is a *unique factorisation domain.*

See also: Euclidean algorithm (p. 202), Euclidean ring (p. 202), Galois field (p. 203), Gaussian field (p. 203), Jacobson's Theorem (p. 206), Unique Factorisation Theorem (p. 213), Wedderburn's Theorem (p. 213).

7 Homomorphisms and quotient algebras

Let $*$ be a binary operation defined on a set X and $\circ$ a second such operation defined on a set Y.

A function $f: X \to Y$ which 'carries' the operation from X to Y, i.e. for which

$$f(a * b) = f(a) \circ f(b) \quad \text{for all} \quad a, b \in X,$$

is called a **homomorphism** (or **morphism**) of $(X, *)$ to $(Y, \circ)$.

In particular, if $(X, *)$ and $(Y, \circ)$ are groups, then f is a **homomorphism of groups.**

$f: R \to R'$ is said to be a **ring-homomorphism** of $(R, +, .)$ to $(R', +, .)$ if for all $a, b \in R$,

$$f(a+b) = f(a)+f(b), \quad f(a.b) = f(a).f(b),$$

and if, in addition, f maps the (multiplicative) identity element of R onto that of R'.

Note. (i) In the case of a group homomorphism it is a consequence of the definition that the identity element of X maps onto the identity element of Y – in the case of rings, which do not have a group structure for multiplication, this additional condition must appear in the definition.

(ii) In general, if the two binary operations $*$ and $\square$ are defined on X and the two operations $\circ$ and $\dagger$ on Y, then $f: X \to Y$ is a homomorphism of $(X, *, \square)$ to $(Y, \circ, \dagger)$ if, for all $a, b \in X$

$$f(a * b) = f(a) \circ f(b), \quad f(a \square b) = f(a) \dagger f(b),$$

with the proviso that in the case of certain structures, e.g., rings, it will be necessary to impose additional conditions.

A homomorphism $f: (X, *) \to (Y, \circ)$ is said to be

a **monomorphism** if f is injective,
an **epimorphism** if f is surjective,
an **isomorphism** if f is bijective.

A homomorphism of $(X, *)$ to itself is called an **endomorphism**; an isomorphism of $(X, *)$ to itself is called an **automorphism.**

Similar descriptions are used for homomorphisms of structures on which two operations are defined. Thus a homomorphism of rings is called a mono-, epi- or iso-morphism according as to whether or not it is in-, sur- or bi-jective.

If an isomorphism f exists mapping $(X, *)$ onto $(Y, \circ)$, then we say that $(X, *)$ and $(Y, \circ)$ are **isomorphic structures** and write

$$(X, *) \cong (Y, \circ)$$

or, when no confusion is likely to arise,

$$X \cong Y.$$

Let $f: G \to H$ be a homomorphism of groups. The **image** of f, denoted by $\mathrm{Im}(f)$, is defined to be $\mathrm{Im}(f) = \{f(g) | g \in G\}$ (cf. p. 13).

The **kernel** of f, $\mathrm{Ker}(f)$, is defined by

$$\mathrm{Ker}(f) = \{g \in G | f(g) = e \in H\}$$

(e denotes the identity element of H).

Note. (i) $\mathrm{Im}(f) \subset H$, $\mathrm{Ker}(f) \subset G$.
(ii) $\mathrm{Im}(f)$ is a *subgroup* of H, $\mathrm{Ker}(f)$ is a *normal subgroup* of G.

If $T \subset H$, then the set

$$\{g \in G | f(g) \in T\}$$

is called the **inverse image** of T (cf. p. 15).

The kernel of a homomorphism of rings $f: R \to R'$, is the set

$$\mathrm{Ker}(f) = \{a \in R | f(a) = 0 \in R'\},$$

i.e., it is the kernel of f considered as a homomorphism of additive groups.

An ordered pair (s, t) of homomorphisms of groups (or *modules*, p. 38)

$$A \xrightarrow{s} B \xrightarrow{t} C$$

is said to be **exact** or **exact at** B when $\mathrm{Im}(s) = \mathrm{Ker}(t)$.

A longer sequence

$$A_0 \xrightarrow{t_1} A_1 \xrightarrow{t_2} A_2 \ldots \xrightarrow{t_n} A_n$$

is said to be **exact** when each pair (t_i, t_{i+1}) is exact at A_i $(i = 1, \ldots, n-1)$.

An exact sequence of the form

$$\{e\} \longrightarrow A \xrightarrow{s} B \xrightarrow{t} C \longrightarrow \{e\}$$

is called a **short exact sequence**.

Given a normal subgroup N (p. 28) of a group we can construct an epimorphism of G having N as its kernel by defining

$$p : g \mapsto Ng$$

where Ng is the coset of N in G (p. 28) containing g.

p maps G onto the quotient set G/N (p. 19) and it can be shown that, when the product of cosets is defined by

$$(Ng)(Nh) = N(gh),$$

G/N is a group, known as the **quotient group** or **factor group** of G by N.

Similarly, if I is a two-sided ideal in a ring R, there is a canonical mapping p which maps every element $r \in R$ onto the coset $I+r$ of the additive subgroup I of R (see note (ii) on p. 28).

If we define the sum and product of cosets by

$$(I+r)+(I+s) = I+(r+s),$$

and
$$(I+r)\,.\,(I+s) = I+(r\,.\,s),$$

then it can be shown that R/I is a ring, the **quotient ring** or **residue class ring** of R by I. p is an epimorphism of rings with kernel I.

Given an integral domain D, the 'smallest' field containing D is known as the **field of quotients** of D and is denoted by $Q(D)$.

Note. The construction of $Q(D)$ from D is analogous to that of $\mathbb{Q}$ from $\mathbb{Z}$ summarised on p. 24.

Examples. The function $f: \mathbb{Z} \to \mathbb{Z}_6$, p. 26, which maps the integer n onto its (positive) remainder when divided by 6 is a group *epimorphism* (*surjective homomorphism*) from $(\mathbb{Z}, +)$ to $(\mathbb{Z}_6, \oplus)$. The function $g: \mathbb{Z} \to \mathbb{Z}_6$ which maps n onto 0 if n is even and onto 3 otherwise is a group *homomorphism* but is *not* an epimorphism. The function $h: \mathbb{Z}_6 \to \mathbb{Z}_6$ which maps 0, 2 and 4 onto 0 and 1, 3 and 5 onto 3 is an *endomorphism* of $(\mathbb{Z}_6, \oplus)$. The *kernel* of f is the *subgroup* $\{..., -12, -6, 0, 6, 12, ...\} \subset \mathbb{Z}$. The *image* of g is the *subgroup* $\{0, 3\} \subset \mathbb{Z}_6$. The *inverse image* of $\{1, 2\}$ under f is the set

$$\{..., -5, -4, 1, 2, 7, 8, ...\} = f^{-1}(\{1, 2\}).$$

The sequence of homomorphisms

$$\{e\} \xrightarrow{o_1} \mathbb{Z}_2 \xrightarrow{s} \mathbb{Z}_6 \xrightarrow{t} \mathbb{Z}_3 \xrightarrow{o_2} \{e\}$$

in which s is defined by $s(0) = 0$, $s(1) = 3$,

and

t is defined by $t(0) = 0$, $t(1) = 1$, $t(2) = 2$, $t(3) = 0$, $t(4) = 1$, $t(5) = 2$, is a *short exact sequence*. We note that, as is always the case, s is a *monomorphism* (*injective homomorphism*) (since $\text{Ker}(s) = \text{Im}(o_1) = \{e\}$), t an *epimorphism* (*surjective homomorphism*) (since $\text{Im}(t) = \text{Ker}(o_2) = \mathbb{Z}_3$) and $\text{Im}(s) = \text{Ker}(t)$.

$N = \{0, 3\}$ is a *normal subgroup* of $\mathbb{Z}_6$ and so we can form the *quotient group* $\mathbb{Z}_6/N$ which comprises the three *cosets* $A = \{0, 3\}$, $B = \{1, 4\}$ and $C = \{2, 5\}$ and which is *isomorphic* to the group $\mathbb{Z}_3$ under the *isomorphism* ϕ defined by

$$\phi(A) = 0, \quad \phi(B) = 1, \quad \phi(C) = 2.$$

If we consider the *rings* $(\mathbb{Z}, +, .)$ and $(\mathbb{Z}_6, \oplus, \odot)$, then it can be shown that f, as defined above, is an *epimorphism of rings* with kernel

$$I = \{\ldots, -12, -6, 0, 6, 12, \ldots\}.$$

I is an *ideal* of $\mathbb{Z}$ and the quotient ring $\mathbb{Z}/I$ is isomorphic to $\mathbb{Z}_6$. Note that g, as defined above, is *not* a *homomorphism of rings* since it does not map the multiplicative identity element of $\mathbb{Z}$ onto that of $\mathbb{Z}_6$.

See also: Cayley's Theorem (p. 200); the Isomorphism Theorems (p. 205).

8 Vector spaces and matrices

Let F be a field. A **vector (or linear) space,** V, over F is defined to be an additive Abelian group V together with a function $F \times V \to V$, $(\lambda, v) \mapsto \lambda v$, satisfying

$$\lambda(a+b) = \lambda a + \lambda b,$$

$$(\lambda + \mu)a = \lambda a + \mu a,$$

$$(\lambda\mu)a = \lambda(\mu a),$$

$$1a = a,$$

for all $\lambda, \mu \in F$ and $a, b \in V$.

If in this definition we relax the conditions by allowing F to be a ring R, then the resulting structure is called a **left R-module.** If, further, the function $(\lambda, v) \mapsto \lambda v$ is replaced by one in which the scalar is written on the right, i.e. $(\lambda, v) \mapsto v\lambda$, then we have a **right R-module.** If R is commutative, then the right and left modules will not differ.

F is called the **ground field** of the vector space, its elements are called **scalars** and those of V are called **vectors.**

Note. (i) The elements of modules are also referred to as vectors. The theories of modules and vector spaces begin to diverge when one considers *bases* and *dimension.*

(ii) Letters representing vectors are often distinguished from those representing scalars by being printed in bold type, and we shall adopt this convention in later sections when we come to consider topics in analysis. However, this convention can lead to some confusion in the study of linear algebra, particularly, say, when we wish to consider the ground field $\mathbb{R}$ as a vector space V in its own right.

A **vector subspace** of a vector space V is defined to be any subset V' of V for which

(*a*) V' is a subgroup of the additive group V,

(*b*) $\forall a \in V'$ and $\forall \lambda \in F$, $\lambda a \in V'$.

Example. The set of all n-tuples $(x_1, \ldots, x_n)$ where $x_i \in \mathbb{R}$, $i = 1, \ldots, n$, forms a *vector space,* which we denote by $\mathbb{R}^n$, over the *ground field* $\mathbb{R}$ when we define $(x_1, \ldots, x_n) + (y_1, \ldots, y_n)$ to be $(x_1 + y_1, \ldots, x_n + y_n)$ and $\lambda(x_1, \ldots, x_n)$, $\lambda \in \mathbb{R}$, to be $(\lambda x_1, \ldots, \lambda x_n)$. A typical element of $\mathbb{R}^n$ is the *vector* $(1, 2, 3, \ldots, n)$. If we make the restriction that $x_i \in \mathbb{Z}$, $i = 1, \ldots, n$, and $\lambda \in \mathbb{Z}$, then we obtain the *module* $\mathbb{Z}^n$ over the *ground ring* $\mathbb{Z}$.

A vector space V over a field F which satisfies the ring axioms in such a way that addition in the ring is addition in the vector space and such that $\lambda(v_1 v_2) = (\lambda v_1) v_2 = v_1(\lambda v_2)$ for all $v_1, v_2 \in V$ and $\lambda \in F$ is said to be an **algebra (*F*-algebra, linear algebra) over *F***. If, in addition, it forms a commutative ring, then we say it is a **commutative algebra.** (See, for example, $M_n(F)$ (p. 42), $F[x]$ (p. 55), the quaternion algebra (an algebra over $\mathbb{R}$) (p. 70) and the algebra $\mathbb{R}^I$ (p. 116).)

A subset of an algebra V is termed a **sub-algebra** if it is both a vector subspace and a subring (p. 30) of V.

Note. The hierarchy of structures (cf. p. 31) is, therefore,

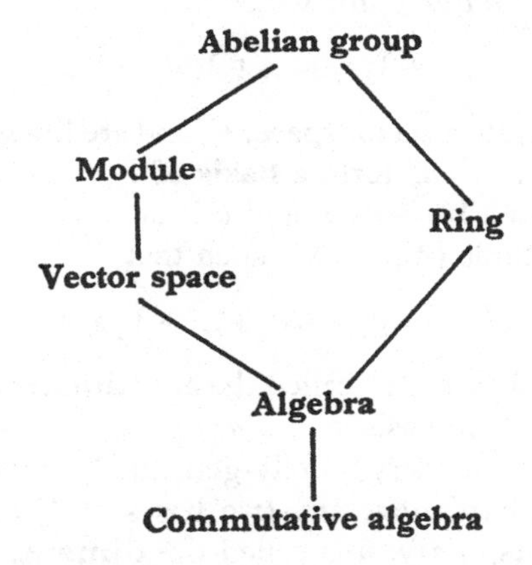

Given two vector spaces U and V over a field F, a **homomorphism** of U to V is a function $t: U \to V$ satisfying

$$t(a+b) = t(a)+t(b),$$

$$t(\lambda a) = \lambda t(a),$$

for all $a, b \in U$ and $\lambda \in F$.

Homomorphisms of vector spaces therefore preserve **linear combinations** of the type

$$\lambda_1 a_1 + \lambda_2 a_2 + \ldots + \lambda_n a_n,$$

where $\lambda_1, \ldots, \lambda_n \in F, \quad a_1, \ldots, a_n \in U$.

For this reason a homomorphism of vector spaces is called a **linear transformation** or **linear mapping.**

The set of all linear combinations of $a_1, \ldots, a_n \in U$ forms a vector subspace of U, called the subspace **generated** or **spanned** by $a_1, \ldots, a_n$.

A vector space, V, is said to be **finitely generated** if there exists a finite set of elements $a_1, \ldots, a_n$ which generate V.

The vectors $a_1, \ldots, a_n$ are said to be **linearly independent** if the only choice of $\lambda_1, \ldots, \lambda_n$ satisfying the relation

$$\lambda_1 a_1 + \lambda_2 a_2 + \ldots + \lambda_n a_n = 0 \quad (0 \in V)$$

is
$$\lambda_1 = \lambda_2 = \ldots = \lambda_n = 0 \quad (0 \in F).$$

The vectors $a_1, \ldots, a_n$ are **linearly dependent** if and only if there exist $\lambda_1, \ldots, \lambda_n$, *not all zero*, for which

$$\lambda_1 a_1 + \lambda_2 a_2 + \ldots + \lambda_n a_n = 0.$$

If $a_1, \ldots, a_n$ generate a vector space, V, and are linearly independent, then we say that $a_1, \ldots, a_n$ form a **basis** of V.

If $a_1, \ldots, a_n$ form a basis of V and b is any vector of V, then there exists a unique n-tuple $(\lambda_1, \ldots, \lambda_n)$ such that

$$b = \lambda_1 a_1 + \lambda_2 a_2 + \ldots + \lambda_n a_n.$$

The scalars $\lambda_1, \ldots, \lambda_n$ are then called the **coordinates** or **components** of b with respect to the basis $a_1, \ldots, a_n$.

It can be shown that every finitely-generated vector space, V, has a basis and that, in particular, any two bases of V contain the same number of elements, n say. n is called the **dimension** of the vector space V over F and we write $\dim V = n$. The vector space V is then isomorphic to F^n, the vector space of all n-tuples $(x_1, \ldots, x_n)$ with $x_i \in F$, $i = 1, \ldots, n$. By definition, $\dim \{0\} = 0$.

Example. The set V' of all vectors belonging to $\mathbb{R}^n$ which can be expressed in the form $\lambda(1, 1, 1, \ldots, 1) + \mu(1, 0, 0, \ldots, 0)$ where $\lambda, \mu \in \mathbb{R}$ constitutes a *vector subspace* of $\mathbb{R}^n$, the subspace *generated* by the vectors $a_1 = (1, 1, \ldots, 1)$ and $a_2 = (1, 0, 0, \ldots, 0)$. Since V' is generated by two vectors it is *finitely generated* and, moreover, a_1 and a_2 form a *basis* for V' since $\lambda_1 a_1 + \lambda_2 a_2 = 0$ implies $\lambda_1 = \lambda_2 = 0$, i.e. a_1 and a_2 are *linearly independent*. The *dimension* of V' over $\mathbb{R}$ is 2 and that of $\mathbb{R}^n$ (over $\mathbb{R}$) is n. V' is *isomorphic* to $\mathbb{R}^2$, i.e. the mapping t which maps a_1 onto $(1, 0)$ and a_2 onto $(0, 1)$ is an *isomorphism* of vector spaces. (We note that $\mathbb{R}^2$ has dimension 2 over $\mathbb{R}$ and dimension 1 over $\mathbb{C}$ (the field of complex numbers, p. 67).) The vector $b = (2, 3, 3, \ldots, 3) \in V'$ since $b = 3a_1 - a_2$ and the *components* or *coordinates* of b with respect to the basis (a_1, a_2) are $(3, -1)$.

If U and V are finite-dimensional vector spaces over the same field F, and if 'addition' of linear transformations and 'multiplication of linear transformations by a scalar' are defined by

$$(t_1+t_2)(a) = t_1(a)+t_2(a),$$
$$(\lambda t)(a) = \lambda(t(a)), \quad \text{for all } a \in U,$$

then it is easily shown that the set of linear transformations itself forms a finite-dimensional vector space over F (having dimension $(\dim U) \times (\dim V)$). This vector space is denoted by $\mathrm{Hom}(U, V)$, $\mathscr{L}_F(U, V)$ or $\mathscr{L}(U, V)$.

Let $t \in \mathrm{Hom}(U, V)$, $u_1, \ldots, u_n$ be a basis for U and $v_1, \ldots, v_m$ be a basis for V. Then t is completely determined by the formulae which tell how the components $(x_1', \ldots, x_m')$ of the vector $t(x)$, with respect to the basis $v_1, \ldots, v_m$ of V, can be obtained in terms of $(x_1, \ldots, x_n)$, the components of the vector x with respect to the basis $u_1, \ldots, u_n$ of U. We have

$$\begin{aligned} x_1' &= a_{11}x_1+a_{12}x_2+\ldots+a_{1n}x_n, \\ x_2' &= a_{21}x_1+a_{22}x_2+\ldots+a_{2n}x_n, \\ &\vdots \\ x_m' &= a_{m1}x_1+a_{m2}x_2+\ldots+a_{mn}x_n, \quad a_{ij} \in F, \end{aligned}$$

and the coefficients a_{ij} determine the homomorphism t uniquely with respect to the chosen bases.

The rectangular array of coefficients

$$\begin{pmatrix} a_{11} & a_{12} & \ldots & a_{1n} \\ a_{21} & a_{22} & \ldots & a_{2n} \\ \vdots & \vdots & & \vdots \\ a_{m1} & a_{m2} & \ldots & a_{mn} \end{pmatrix}$$

is said to form a **matrix**, A, having m **rows** and n **columns.**

Note. More abstractly we can think of the matrix as a function

$$A : \{1, 2, \ldots, m\} \times \{1, 2, \ldots, n\} \to F.$$

(Compare the abstract definition of a *sequence* (p. 108).)

The matrix A is often abbreviated to (a_{ij}), and to denote that this represents the linear transformation t we write $(t) = (a_{ij})$ or, where there is ambiguity concerning the choice of bases, $(t; u_k, v_l) = (a_{ij})$.

The matrix corresponding to the *zero mapping*, $t : x \mapsto 0$ (all $x \in U$), is called the **zero matrix** and is denoted by 0.

Example. The *transformation* $t: V' \to \mathbb{R}^2$ defined on p. 40 is described *with respect to the bases* (a_1, a_2) of V' and $((1, 0), (0, 1))$ of $\mathbb{R}^2$ by the *matrix*

$$\begin{pmatrix} 1 & 0 \\ 0 & 1 \end{pmatrix}$$

since, for example, the vector b with components $(3, -1)$ is mapped onto $t(b) = (3, -1)$ which has components $(3, -1)$ relative to our basis in $\mathbb{R}^2$. If, however, we take $(2, 0)$ and $(1, 3)$ as the basis for $\mathbb{R}^2$, then we should have

$$x_1' = \tfrac{1}{2}x_1 - \tfrac{1}{6}x_2,$$
$$x_2' = 0x_1 + \tfrac{1}{3}x_2,$$

i.e., the components of b relative to the new basis would be $(\frac{5}{3}, -\frac{1}{3})$, and t would be described by the matrix

$$\begin{pmatrix} \frac{1}{2} & -\frac{1}{6} \\ 0 & \frac{1}{3} \end{pmatrix}.$$

The **sum** of two $m \times n$ matrices, $A = (a_{ij})$ and $B = (b_{ij})$, is defined to be the $m \times n$ matrix $A + B = (a_{ij} + b_{ij})$.

The **product** of the $m \times n$ matrix $A = (a_{ij})$ **with the scalar** λ is defined to be the $m \times n$ matrix $\lambda A = (\lambda a_{ij})$.

With these definitions the set of all $m \times n$ matrices with coefficients in F becomes a vector space of dimension mn and is isomorphic to $\mathrm{Hom}(U, V)$.

We define the **product matrix,** $C = AB$, of two matrices $A = (a_{ij})$ an $m \times n$ matrix, and $B = (b_{jk})$ an $n \times p$ matrix, to be the $m \times p$ matrix (c_{ik}) where

$$c_{ik} = \sum_{j=1}^{n} a_{ij} b_{jk}.$$

This definition is motivated by the need to form the composite linear transformation $s \circ t \in \mathrm{Hom}(U, W)$, given $t \in \mathrm{Hom}(U, V)$ corresponding to the matrix B, and $s \in \mathrm{Hom}(V, W)$ corresponding to the matrix A. The matrix product AB of the matrices A and B is not defined unless the number of columns of A is equal to the number of rows of B (is equal to the dimension of V). The existence of the product AB will not therefore imply the existence of the product BA.

With this definition of multiplication, the set of **square** matrices of order n, i.e. those matrices having n rows *and* n columns, with coefficients in a *ring* K, form a *non-commutative ring* (provided $n > 1$ and $K \neq \{0\}$) denoted by $M_n(K)$.

A matrix $A \in M_n(K)$ is said to be **non-singular** or **invertible in** $M_n(K)$ if there exists $B \in M_n(K)$ such that

$$AB = BA = I_n,$$

where I_n is the **identity** or **unit matrix of order** n having coefficients (δ_{ij}) where δ_{ij} – the **Kronecker delta** – is defined by

$$\delta_{ij} = 0 \quad (i \neq j),$$
$$\delta_{ij} = 1 \quad (i = j).$$

The matrix B is then unique and is known as the **inverse** of A; it is denoted by A^{-1}.

If there is no matrix B in $M_n(K)$ such that $AB = BA = I$, then A is said to be **singular** or **non-invertible** in $M_n(K)$.

Note. A linear mapping $U \to V$ will be an isomorphism of vector spaces if and only if it can be represented by an invertible (square) matrix.

The **transpose** of the $m \times n$ matrix $A = (a_{ij})$ is defined to be the $n \times m$ matrix (a_{ji}) obtained from A by interchanging rows and columns, i.e. the element appearing in the ith row and jth column of A will appear in the jth row and ith column of the transpose of A. The transpose of A is denoted by A^T, A^t, ${}^tA$ or A'.

When $A = A^T$ the matrix A is said to be **symmetric**.

Example. If

$$A = \begin{pmatrix} 2 & 3 & 2 \\ 1 & 4 & 1 \end{pmatrix} \quad \text{and} \quad B = \begin{pmatrix} 1 & 2 \\ 3 & 1 \\ 0 & 1 \end{pmatrix},$$

then

$$C = AB = \begin{pmatrix} 11 & 9 \\ 13 & 7 \end{pmatrix}.$$

The matrix

$$D = \begin{pmatrix} 1 & -3 \\ 2 & 4 \end{pmatrix} \in M_2(\mathbb{Q})$$

is *non-singular* (*invertible*) in $M_2(\mathbb{Q})$, for setting

$$E = \begin{pmatrix} \frac{4}{10} & \frac{3}{10} \\ -\frac{2}{10} & \frac{1}{10} \end{pmatrix}$$

we have

$$ED = DE = \begin{pmatrix} 1 & 0 \\ 0 & 1 \end{pmatrix} = I_2.$$

E is, therefore, the *inverse* of D in $M_2(\mathbb{Q})$ (and, *a fortiori*, in any ring $M_2(R)$ such that $\mathbb{Q}$ is a subring of R). (Since the inverse of D is uniquely defined in $M_2(\mathbb{Q})$ and since $E \notin M_2(\mathbb{Z})$, it follows, however, that D is *singular* (*non-invertible*) in $M_2(\mathbb{Z})$.) The *transposes* of A and D are the matrices

$$A^T = \begin{pmatrix} 2 & 1 \\ 3 & 4 \\ 2 & 1 \end{pmatrix} \quad \text{and} \quad D^T = \begin{pmatrix} 1 & 2 \\ -3 & 4 \end{pmatrix}.$$

The matrix

$$F = \begin{pmatrix} 1 & 3 \\ 3 & 2 \end{pmatrix}$$

is *symmetric*.

For each $n > 0$ the set of non-singular $n \times n$ matrices over the field F forms a multiplicative group, called the **general linear group** $GL(n, F)$.

The elements of $GL(n, F)$ having determinant 1 (p. 52) form a subgroup denoted by $SL(n, F)$ and known as the **special linear group.**

An $m \times 1$ matrix having only one column is known as a **column matrix** or **column vector**, a $1 \times n$ matrix is known as a **row matrix** or **row vector.**

Note. The equations which determine the homomorphism t with matrix A (p. 41) can therefore be written

$$X' = AX,$$

where X' is a column vector whose m elements are the components of $t(x)$ with respect to $v_1, \ldots, v_m$ and X is a column vector having n elements, the components of x with respect to $u_1, \ldots, u_n$.

Thus, in the case of the transformation $t: V' \to \mathbb{R}^2$ defined on p. 42 we have

$$\begin{pmatrix} x_1' \\ x_2' \end{pmatrix} = \begin{pmatrix} \frac{1}{2} & -\frac{1}{6} \\ 0 & \frac{1}{3} \end{pmatrix} \begin{pmatrix} x_1 \\ x_2 \end{pmatrix},$$

where x_1, x_2 are the components of $x \in V'$ with respect to the basis (a_1, a_2) and x_1', x_2' are the components of $t(x) \in \mathbb{R}^2$ with respect to the basis $((2,0), (1,3))$.

In particular, a $1 \times n$ matrix will describe a homomorphism from an n-dimensional vector space U with basis $u_1, \ldots, u_n$ to a one-dimensional vector space V with basis v_1. The set of all $1 \times n$ matrices will form a vector space isomorphic to the vector space $\mathrm{Hom}(U, F)$ of all homomorphisms mapping U onto its ground field F, which can be regarded as a vector space of dimension 1 over itself.

In general, if V is any vector space over a field F, then the vector space $\mathrm{Hom}(V, F)$ is called the **dual space** of V and is denoted by V^* or $\hat{V}$. Elements of $\mathrm{Hom}(V, F)$ are known as **linear functionals, linear forms** or **covectors.**

Note. It can be shown that every finite-dimensional vector space is isomorphic to its dual.

Let S be a subspace of the vector space V. Consider the set $S^\circ \subset V^*$ of all linear forms f on V which satisfy

$$f(a) = 0 \quad \text{for all} \quad a \in S.$$

The set S° is a subspace of V^*, called the **annihilator** of S in V^*, and the spaces S and S° are said to be **orthogonal.**

Note. It follows that $\dim S + \dim(\text{annihilator } S) = \dim V$.

Example. A *homomorphism* (*linear functional*) $t: V(=\mathbb{R}^n) \to \mathbb{R}$ will be of the form

$$t: (x_1, x_2, \ldots, x_n) \mapsto a_{11}x_1 + \ldots + a_{1n}x_n, \quad a_{1i} \in \mathbb{R}, i = 1, \ldots, n.$$

t will, therefore, be described relative to the basis $(e_1, \ldots, e_n)$ of V, where $e_i = (\delta_{1i}, \delta_{2i}, \ldots, \delta_{ni})$, $i = 1, \ldots, n$ (see p. 43), and to the basis (1) of $\mathbb{R}$ by the n-tuple of real numbers (i.e., $1 \times n$ matrix with real coefficients) $(a_{11}, \ldots, a_{1n})$. The set of all such n-tuples (each one of which represents a well-defined *linear functional*) forms a vector space V^*, the *dual space* of V, which is isomorphic to V. If V' is the subspace of V defined on p. 40, then the *annihilator* of V' in V^* is the subspace of V^* comprising those linear functionals $f \in V^*$ which satisfy

$$f(a_1) = f(a_2) = 0.$$

This subspace will have dimension $n-2$.

Given two subspaces S and T of a vector space V, we denote by $S+T$ the set $\{x+y \mid x \in S, y \in T\}$. It is easy to verify that $S+T$ is a subspace of V. We say that V is the **direct sum** of S and T, written $S \oplus T$, if and only if $V = S+T$ and $S \cap T = \{0\}$. S and T are then called **direct summands** of V.

Note. (i) Any subspace S of a finite-dimensional vector space V is a *direct summand* of V. Moreover, if $V = S \oplus T$, then

$$\dim T = \dim V - \dim S.$$

(ii) If S and T are any finite-dimensional subspaces of a vector space V, then

$$\dim S + \dim T = \dim(S \cap T) + \dim(S+T).$$

Example. Let V' be defined as on p. 40. Then $\mathbb{R}^n = V' \oplus S$ where S is the subspace of $\mathbb{R}^n$ with basis $(0, 1, 0, \ldots, 0), (0, 0, 1, 0, \ldots, 0), \ldots, (0, 0, \ldots, 0, 1, 0)$. If T denotes the subspace of $\mathbb{R}^n$ generated by $(0, 1, 0, \ldots, 0)$ and $(1, 1, \ldots, 1)$, then $\dim S = n-2$, $\dim T = 2$, $\dim V' = 2$, $\dim(S \cap T) = 1$, $\dim(S+T) = n-1$, $\dim(T \cap V') = 1$, $\dim(T+V') = 3$.

9 Linear equations and rank

Linear equations

If F is a field, then a **system of m linear equations** in n unknowns with coefficients in F is any system of relations

$$\begin{array}{ccc} a_{11}x_1 + a_{12}x_2 + \ldots + a_{1n}x_n & = & b_1, \\ \vdots & & \vdots \\ a_{m1}x_1 + a_{m2}x_2 + \ldots + a_{mn}x_n & = & b_m, \end{array}$$

in which the **coefficients** a_{ij} and the **constant terms** b_i are fixed elements of F.

The system is said to be **homogeneous** when $b_1 = b_2 = \ldots = b_m = 0$.

A **solution** of the system (in the field F) is any vector $x = (x_1, \ldots, x_n)$ whose components satisfy all the relations.

The definitions given above can be generalised by replacing the field F by a ring K. We cannot then, however, make use of the properties of vector spaces in the ensuing theory. Equations (not necessarily linear) over the ring $\mathbb{Z}$ are historically known as **Diophantine equations.**

The system of equations can be rewritten in the form

$$Ax = b,$$

where A is the $m \times n$ matrix with coefficients a_{ij} and x and b are, respectively, the column matrices

$$x = \begin{pmatrix} x_1 \\ \vdots \\ x_n \end{pmatrix}, \qquad b = \begin{pmatrix} b_1 \\ \vdots \\ b_m \end{pmatrix},$$

having n and m rows respectively.

Finding the solution of the system of equations

$$Ax = b$$

can be interpreted as finding the inverse image of b under the vector space homomorphism t determined by A which maps the n-dimensional vector space F^n (p. 40) into the m-dimensional space F^m.

The **rank**, r, of the homomorphism t is defined to be the dimension of the subspace of F^m onto which F^n is mapped by t, i.e.

$$r = \dim(\mathrm{Im}(t)) = \dim(t(F^n)).$$

This rank will be determined by the number of linearly independent columns of A, i.e. it will be the dimension of the subspace of F^m spanned by the n vectors

$$\begin{pmatrix} a_{11} \\ a_{21} \\ \vdots \\ a_{m1} \end{pmatrix}, \begin{pmatrix} a_{12} \\ a_{22} \\ \vdots \\ a_{m2} \end{pmatrix}, \ldots, \begin{pmatrix} a_{1n} \\ a_{2n} \\ \vdots \\ a_{mn} \end{pmatrix}.$$

It is known as the **column rank of the matrix** A.

The dimension of the subspace of $(F^n)^*$ spanned by the m linear forms

$$\begin{matrix} (a_{11}, & a_{12}, & \ldots, & a_{1n}), \\ \vdots & \vdots & & \vdots \\ (a_{m1}, & a_{m2}, & \ldots, & a_{mn}), \end{matrix}$$

is called the **row rank of the matrix** A.

Note. (i) Consideration of the system $Ax = 0$ and use of the result on the dimension of the annihilator (p. 44) show that the row rank and the column rank of a matrix are equal.

(ii) Alternatively (and equivalently) the **rank** of an $m \times n$ matrix A can be defined to be the largest integer r such that a non-singular (and so square) matrix of order r can be obtained from A by removing $m-r$ of its rows and $n-r$ of its columns. (Compare the definition of a minor on p. 52.)

In the case of a homogeneous system, $Ax = 0$, the inverse image of 0, i.e. the set of solutions, is $\operatorname{Ker}(t)$ (p. 35) which is a subspace of F^n (known as the **null space** of t) of dimension $n-r$, a number known as the **nullity** of t (or of A).

In the non-homogeneous case, if a is any one solution of the system, known as a **particular solution**, then the set of solutions is the coset $a + \operatorname{Ker}(t)$. This coset is not a subspace of F^n (unless $a = 0$) and is often referred to as a **flat**.

Example. Consider the *system of 3 equations in 4 unknowns*:

$$x_1 + x_3 = 1,$$

$$x_2 + x_3 + 2x_4 = 0,$$

$$x_1 + x_2 + 2x_3 + 2x_4 = 1,$$

having *coefficients* and *constant terms* in the field $\mathbb{Q}$. The system can be rewritten in the form

$$Ax = b,$$

where A is the 3×4 matrix $\begin{pmatrix} 1 & 0 & 1 & 0 \\ 0 & 1 & 1 & 2 \\ 1 & 1 & 2 & 2 \end{pmatrix}$

and x and b are respectively the column matrices

$$\begin{pmatrix} x_1 \\ x_2 \\ x_3 \\ x_4 \end{pmatrix} \quad \text{and} \quad \begin{pmatrix} 1 \\ 0 \\ 1 \end{pmatrix}.$$

Since b has a non-zero element, the system is *non-homogeneous*. A *particular solution*, i.e., one solution, of the system in $\mathbb{Q}$ is $(1, -2, 0, 1)$ since this vector has components which satisfy all the equations.

The matrix A determines a unique linear mapping t from $\mathbb{Q}^4$ to $\mathbb{Q}^3$ (relative to pre-assigned bases). The solution set in $\mathbb{Q}$ of the system will be the *inverse image* of b under t. The *rank* of t is the dimension of the subspace of $\mathbb{Q}^3$ generated by the vectors

$$\begin{pmatrix} 1 \\ 0 \\ 1 \end{pmatrix}, \begin{pmatrix} 0 \\ 1 \\ 1 \end{pmatrix}, \begin{pmatrix} 1 \\ 1 \\ 2 \end{pmatrix}, \begin{pmatrix} 0 \\ 2 \\ 2 \end{pmatrix}$$

and can be shown to be 2. We note also that (i) striking out the 3rd row and the 3rd and 4th columns of A we obtain a non-singular matrix of order 2 and that it is impossible to obtain a matrix of order 3 in like manner, i.e., rank $A = 2$, and (ii) the subspace of $(\mathbb{Q}^4)^*$ spanned by $(1, 0, 1, 0)$, $(0, 1, 1, 2)$ and $(1, 1, 2, 2)$ has also dimension 2.

$\operatorname{Ker}(t)$ is a subspace of $\mathbb{Q}^4$ of dimension $4-2 = 2$ (the *nullity* of A) and is generated by the vectors $(1, 0, -1, \frac{1}{2})$ and $(0, 1, 0, -\frac{1}{2})$, i.e. it comprises those vectors of $\mathbb{Q}^4$ of the form $(s, t, -s, \frac{1}{2}(s-t))$, where $s, t \in \mathbb{Q}$. ($\operatorname{Ker}(t)$ is the set of solutions in $\mathbb{Q}$ of the homogeneous system $Ax = 0$.)

The set of solutions in $\mathbb{Q}$ of the system $Ax = b$ is obtained by considering $\operatorname{Ker}(t)$ and a *particular solution*, and is the *flat*

$$(1, -2, 0, 1) + (s, t, -s, \tfrac{1}{2}(s-t)), \quad s, t \in \mathbb{Q}.$$

Thus the solution of the system is

$$x_1 = 1+s, \quad x_2 = -2+t, \quad x_3 = -s, \quad x_4 = 1+\tfrac{1}{2}(s-t)$$

where $s, t \in \mathbb{Q}$. (Note that by taking $s, t \in \mathbb{R}$ we obtain the set of solutions of the system in the field $\mathbb{R}$.)

A system of equations for which $m = n = r$ is known as a **Cramer** or **regular system.** It has the unique solution $x = A^{-1}b$.

Elementary operations and special types of matrices

In order to find the rank of a matrix it is often necessary explicitly to test whether or not a given list of vectors is linearly independent. To do so we make use of the **elementary operations** which transform one set of generators of a subspace into another (hopefully, more simple) set.

The operations are
(*a*) interchanging two vectors in the list,
(*b*) multiplying one vector of the list by a non-zero scalar,
(*c*) adding a scalar multiple of one vector of the list to another.
Substituting 'row' or 'column' for 'vector' we obtain a description of the *elementary operations* that transform a given matrix to a matrix having equal rank.

An **elementary matrix** is any square matrix obtained by applying one of the elementary operations to the columns of an identity matrix, e.g.

$$\begin{pmatrix} 0 & 1 & 0 & 0 & 0 & \dots & 0 \\ 1 & 0 & 0 & 0 & 0 & \dots & 0 \\ 0 & 0 & 1 & 0 & 0 & \dots & 0 \\ \vdots & \vdots & \vdots & \vdots & \vdots & & \vdots \\ 0 & 0 & 0 & 0 & 0 & \dots & 1 \end{pmatrix}.$$

A matrix is said to be in **echelon form** when its first non-zero row is $(1, 0, 0, \dots, 0)$ and when that portion of the matrix below that row and omitting the first column is in echelon form, e.g.

$$\begin{pmatrix} 0 & 0 & 0 \\ 1 & 0 & 0 \\ a_1 & 0 & 0 \\ a_2 & 1 & 0 \\ a_3 & a_4 & 1 \end{pmatrix} \quad \text{(here } a_1, \dots, a_4 \text{ are arbitrary elements of } F\text{).}$$

Applying further elementary operations one can make all the a_i zero, in which case we say the matrix is in **reduced echelon form.**

Additionally, A may be described as in *echelon form* if its transpose, A^T, is in echelon form.

A matrix $A = (a_{ij})$ is said to be in **triangular form** if either

$$(a)\ a_{ij} = 0 \quad \text{all} \quad i > j,$$

or

$$(b)\ a_{ij} = 0 \quad \text{all} \quad i < j.$$

A is said to be in **diagonal form** if $a_{ij} = 0,\ i \neq j$.

Two square matrices A and B of order n over the same field F are said to be **similar** (over F) if and only if there is a non-singular matrix $P \in M_n(F)$ for which

$$B = PAP^{-1}.$$

Two $m \times n$ matrices A and B are **equivalent** over F if and only if there are non-singular square matrices P and Q over F for which

$$B = QAP^{-1}.$$

Note. (i) Two similar matrices are, *a fortiori*, equivalent.

(ii) If A and B both represent the same transformation (but with respect to different pairs of bases), then they are *equivalent*.

(iii) It can be shown that any $m \times n$ matrix of rank r is equivalent over F to a matrix $D_{(r)}^{(m\times n)} = (d_{ij})$ given by

$$d_{ii} = 1 \quad (i = 1, \ldots, r),$$
$$d_{ij} = 0 \quad \text{otherwise.}$$

$D_{(r)}^{(m\times n)}$ is known as a **canonical** or **normal form.**

Example.

Taking
$$A = \begin{pmatrix} 5 & 3 & 2 \\ 1 & 2 & 6 \\ 7 & 0 & -14 \end{pmatrix}, \quad Q = \begin{pmatrix} 2 & -3 & 0 \\ 0 & 1 & 0 \\ -2 & 3 & 1 \end{pmatrix},$$
$$P^{-1} = \begin{pmatrix} \frac{1}{7} & 0 & -\frac{1}{4} \\ -\frac{1}{14} & \frac{1}{2} & \frac{1}{2} \\ 0 & 0 & -\frac{1}{8} \end{pmatrix},$$
we have
$$QAP^{-1} = \begin{pmatrix} 1 & 0 & 0 \\ 0 & 1 & 0 \\ 0 & 0 & 0 \end{pmatrix},$$

i.e. A is *equivalent* (over $\mathbb{Q}$) to the normal form $D_{(2)}^{(3\times 3)}$.

The matrix given on p. 49 as an example of an elementary matrix may be written in the form

$$\left(\begin{array}{c|c} J & O_1 \\ \hline O_2 & I_{n-2} \end{array}\right),$$

where J denotes the 2×2 matrix $\begin{pmatrix} 0 & 1 \\ 1 & 0 \end{pmatrix}$,

O_1 denotes the $2 \times (n-2)$ matrix $\begin{pmatrix} 0 & 0 & \ldots & 0 \\ 0 & 0 & \ldots & 0 \end{pmatrix}$,

O_2 denotes the $(n-2) \times 2$ matrix $\begin{pmatrix} 0 & 0 \\ 0 & 0 \\ \vdots & \vdots \\ 0 & 0 \end{pmatrix}$,

and I_{n-2} denotes the identity matrix of order $n-2$.

The matrix is then said to be **partitioned** into the four **submatrices** J, O_1, O_2 and I_{n-2}.

10 Determinants and multilinear mappings

Permutations and determinants

Consider the *permutation* (bijection) of $\{1, 2, 3, 4\}$ onto itself defined by $1 \mapsto 2$, $2 \mapsto 4$, $3 \mapsto 1$, $4 \mapsto 3$. This particular mapping is denoted by

$$\downarrow \begin{pmatrix} 1 & 2 & 3 & 4 \\ 2 & 4 & 1 & 3 \end{pmatrix} \quad \text{or} \quad (1243)$$

and is known as a **cyclic permutation** or a **cycle of length** 4.

Note. (i) It is usual to omit the mapping arrow in the first representation.

(ii) In the second representation each element is followed by its image under the mapping with the exception of the last element which is mapped onto the first. It follows that (1243) and (2431) represent the same permutation.

It can be shown that any permutation can be written as a product of distinct cycles, e.g.

$$\begin{pmatrix} 1 & 2 & 3 & 4 \\ 3 & 4 & 1 & 2 \end{pmatrix} \quad \text{as} \quad (13)\,(24),$$

and that any product of distinct cycles represents a permutation.

Note.

$$\begin{pmatrix} 1 & 2 & 3 & 4 \\ 3 & 1 & 2 & 4 \end{pmatrix}$$

is written as (132) (4) or as (132). If the latter representation is used then there is an especial need to keep the domain of the permutation in mind.

A cycle of length 2 is called a **transposition**. Since any cycle can be shown to be a product of transpositions, e.g.

$$(1\;2\;3\;4) = (14)\,(13)\,(12),$$

it follows that any permutation can be expressed as a product of transpositions.

The representation of a permutation as a product of transpositions will not be unique. However, for a given permutation σ, either all products will be formed of an even number of transpositions or all products will be formed of an odd number. If the former is the case we say that σ is an **even** permutation, if the latter, that it is an **odd** permutation. The **parity** or **sign** of a permutation σ is defined to be $+1$ if σ is even and -1 if σ is odd.

Note. The mapping $\sigma \mapsto \text{sign}(\sigma)$ is a homomorphism of groups

$$S_n \text{ (p. 27)} \to (\{+1, -1\}, \times).$$

When a permutation is represented in the form

$$\begin{pmatrix} 1 & 2 & 3 & 4 & \dots \\ a_1 & a_2 & a_3 & a_4 & \dots \end{pmatrix}$$

and two numbers, say a_3 and a_4, are out of their natural order, we say that such a derangement is an **inversion**, thus

$$\begin{pmatrix} 1 & 2 & 3 & 4 \\ 1 & 2 & 4 & 3 \end{pmatrix}$$

contains the inversion 43.

The number of inversions in any given permutation can be counted and will be even if the permutation is even and odd if the permutation is odd. This result is the basis of an alternative definition of *odd* or *even permutations*.

For each square matrix $A = (a_{ij})$ of order n we now define the **determinant** of A, denoted by $\det A$ or $|A|$, to be the number

$$\Sigma \,\text{sign}(\alpha, \beta, \dots, \nu) a_{1\alpha} a_{2\beta} \dots a_{n\nu},$$

the summation being extended over all $n!$ permutations

$$\begin{pmatrix} 1 & 2 & 3 & \dots & n \\ \alpha & \beta & \gamma & \dots & \nu \end{pmatrix}$$

of the column suffixes.

The coefficient or **cofactor** of a_{ij} in this sum is the determinant of the matrix of order $n-1$ obtained by suppressing the ith row and jth column of A, with the sign $(-)^{i+j}$. This cofactor of a_{ij} in $|A|$ is denoted by $|A_{ij}|$. The matrix obtained by suppressing the ith row and the jth column of A is called the **minor** of a_{ij} in A. (Note that the minor is *not* given the sign $(-)^{i+j}$.) The determinant of the minor of a_{ij} is called the **complementary minor** of a_{ij}.

$|A|$ can then be expressed in terms of the elements of a particular row (or column) together with their cofactors, e.g.,

$$|A| = a_{1i}|A_{1i}| + a_{2i}|A_{2i}| + \dots + a_{ni}|A_{ni}|.$$

This is known as the **expansion of $|A|$ by its ith column.**

Example. The determinant of the 3×3 matrix $A = (a_{ij})$ is

$$|A| = a_{11}a_{22}a_{33} + a_{13}a_{21}a_{32} + a_{12}a_{23}a_{31} - a_{11}a_{23}a_{32} - a_{12}a_{21}a_{33} - a_{13}a_{22}a_{31},$$

since (123), (312) and (231) are *even* permutations with *sign* $+1$ and (132), (213) and (321) are *odd* permutations with *sign* -1.

The *cofactor* of a_{23}, written $|A_{23}|$, is $a_{12}a_{31} - a_{11}a_{32}$.

The *minor* of a_{23} is $\begin{pmatrix} a_{11} & a_{12} \\ a_{31} & a_{32} \end{pmatrix}$.

We note that the cofactor of a_{23} is not equal to the determinant of its minor but to that determinant multiplied by $(-1)^{2+3}$.

The expansion of $|A|$ by its second row yields

$$\begin{aligned} |A| &= a_{21}|A_{21}| + a_{22}|A_{22}| + a_{23}|A_{23}| \\ &= a_{21}(a_{13}a_{32} - a_{12}a_{33}) + a_{22}(a_{11}a_{33} - a_{13}a_{31}) + a_{23}(a_{12}a_{31} - a_{11}a_{32}). \end{aligned}$$

Note. (i) The following rules for the evaluation of determinants are simple consequences of the definition:

(*a*) If two rows (or two columns) of a matrix A are interchanged so as to form the matrix B, then $|A| = -|B|$.

(*b*) If two rows (or columns) of A are identical, then $|A| = 0$.

(*c*) If the matrix B is obtained by adding a constant multiple of any row (column) of A to any other row (column) of A, then $|A| = |B|$.

(*d*) If B is obtained from A by multiplying all the elements in a given row (column) of A by a scalar k, then

$$|B| = k|A|.$$

(ii) It can be shown that if A is a *non-singular* $n \times n$ matrix then the *inverse* of A is the matrix $(1/|A|)\alpha$, where α, the **adjugate** matrix to A, is the matrix obtained by transposing the matrix of *cofactors* of A, i.e. if $A = (a_{ij})$, then

$$\alpha = (|A_{ij}|)^T.$$

(iii) When t, a linear transformation of a vector space V to itself, is represented by a matrix A, it can be shown that $|A|$ is independent of the choice of basis and can therefore be referred to without ambiguity as the **determinant of the linear transformation** t.

Multilinear mappings

An alternative, more abstract, definition of a determinant is a consequence of the following sequence of definitions.

Let $X_1, \ldots, X_n$ and V be vector spaces over a field F. A mapping of the form

$$f: X_1 \times X_2 \times \ldots \times X_n \to V$$

is said to be **multilinear** (n-linear) if for each $i = 1, \ldots, n$ and for all choices of vectors $a_1 \in X_1, a_2 \in X_2, \ldots, a_n \in X_n$, the mapping

$$x \mapsto f(a_1, a_2, \ldots, a_{i-1}, x, a_{i+1}, \ldots, a_n)$$

of X_i into V is linear.

Note. This definition can be extended to the case where $X_1, \ldots, X_n$ and V are modules over a commutative ring.

In particular, $f: X \times Y \to V$ is **bilinear** if

$$f(x_1, \lambda y_1 + \mu y_2) = \lambda f(x_1, y_1) + \mu f(x_1, y_2)$$

and
$$f(\lambda x_1 + \mu x_2, y_1) = \lambda f(x_1, y_1) + \mu f(x_2, y_1),$$

for all $x_1, x_2 \in X$, $y_1, y_2 \in Y$ and $\lambda, \mu \in F$.

An **n-linear form on X** (a vector space over the field F) is an n-linear mapping of X^n into F.

A **tensor** of type $\binom{m}{n}$ is an $(m+n)$-linear mapping of $(X^*)^m \times X^n$ into F. (Here X^* denotes the dual space (p. 44).) m is called the **covariant** index and n the **contravariant** index of the tensor.

A bilinear mapping $f: X \times X \to V$ is said to be **alternating** if $f(x, x) = 0$ for all $x \in X$.

An n-linear mapping $f: X^n \to V$ is said to be **alternating** if $f(x_1, x_2, \ldots, x_n) = 0$ whenever any two of the $x_1, \ldots, x_n$ are equal.

The mapping $(A_1, \ldots, A_n) \mapsto |A|$ where $A_1, \ldots, A_n$ are the columns of the matrix A of order n is then an alternating, n-linear form from

$$\underbrace{F^n \times F^n \times \ldots \times F^n}_{n \text{ times}}$$

into F. It can be shown that there is only one such mapping of $(A_1, \ldots, A_n)$ to F which maps the identity matrix onto the unit element. The image of A under this function can therefore be *defined* to be $|A|$.

Examples. (i) Consider $f: \mathbb{R} \times \mathbb{R} \times \ldots \times \mathbb{R} \to \mathbb{R}$, defined by

$$f(x_1, \ldots, x_n) = x_1 x_2 \ldots x_n.$$

f is *multilinear* (n-linear).

(ii) Consider $f: \mathbb{R}^2 \times \mathbb{R}^2 \to \mathbb{R}$, defined by $f(x, y) = x.y$ where $x.y$ denotes the scalar product (p. 169) of the vectors x and y (i.e. $x.y = x_1 y_1 + x_2 y_2$). f is *bilinear* on $\mathbb{R}^2 \times \mathbb{R}^2$.

(iii) Consider $f: \mathbb{R}^3 \times \mathbb{R}^3 \to \mathbb{R}^3$, defined by $f(x, y) = x \times y$ where $x \times y$ denotes the vector product (p. 169) of the vectors x and y. Then f is an *alternating, bilinear mapping*.

(iv) A linear form on X (p. 44) is a *tensor* of type $\binom{0}{1}$.

11 Polynomials

Consider the set $K^{\mathbb{N}}$ (p. 13) where K is a commutative ring. An element of $K^{\mathbb{N}}$ is a mapping $f: \mathbb{N} \to K$ which can be described by listing the images of 0, 1, 2, ... in the manner $(f_0, f_1, f_2, \ldots)$, $f_i \in K$. Such a map is known as a **sequence** or a **sequence of elements of** K.

With any sequence of $K^{\mathbb{N}}$ we can associate the **formal power series**

$$f_0+f_1x+f_2x^2+\ldots+f_nx^n+\ldots.$$

Moreover, we can define the 'sum' and 'product' of two sequences (formal power series) f and g to be

$$f+g: \mathbb{N} \to K,$$

where $\quad (f+g)_n = f_n+g_n, \quad n = 0, 1, 2, \ldots,$

and $\quad f.g: \mathbb{N} \to K,$

where $\quad (f.g)_n = f_0g_n+f_1g_{n-1}+\ldots+f_{n-1}g_1+f_ng_0.$

With these definitions $K^{\mathbb{N}}$ is a commutative ring with multiplicative identity (1, 0, 0, ...).

The particular sequence (0, 1, 0, 0, ...), corresponding to the formal power series x, is called the **indeterminate** x and the **ring,** $K^{\mathbb{N}}$**, of all formal power series** over K is denoted by $K[[x]]$.

A **polynomial** or **polynomial form** is defined to be a formal power series having only a finite number of non-zero coefficients, i.e. it is a sequence $f: \mathbb{N} \to K$ for which there exists an m such that $f(n) = 0$ for all $n > m$. The smallest such number m $(= N)$ is then called the **degree** of the polynomial f. The degree of the zero polynomial is either left undefined or is said to be $-\infty$ (p. 67). A polynomial of degree 0 is called a **constant polynomial** while those of degree 1, 2, 3, 4, 5, ... are said to be **linear, quadratic, cubic, quartic, quintic,** ....

If f is a polynomial of degree N and the **leading coefficient,** f_N, equals 1, then f is said to be **monic.**

The polynomials form a subring of $K[[x]]$ called the **ring of polynomials over** K and denoted by $K[x]$.

The polynomial $(1, 5, 3, 7, 0, 0, \ldots) \in \mathbb{Z}[x]$ will then be written in terms of the indeterminate x as $1+5x+3x^2+7x^3$.

Since $K[x]$ is a commutative ring we can repeat the arguments given above and form the ring of polynomials over $K[x]$ with indeterminate y. Our notation would lead us to describe this latter ring as $(K[x])[y]$ or $K[x][y]$. Without any loss of precision we can denote the ring by $K[x, y]$: it is the **polynomial ring in the two indeterminates** x and y, and consists of all finite sums of the type

$$\Sigma a_{\alpha\beta}x^{\alpha}y^{\beta}, \quad \alpha, \beta \in \mathbb{N}, \quad a_{\alpha\beta} \in K.$$

In general, the **polynomial ring in the n indeterminates** $x_1, \ldots, x_n$ is denoted by $K[x_1, \ldots, x_n]$, and it consists of all the finite sums

$$\Sigma a_{\alpha_1\alpha_2\ldots\alpha_n} x_1^{\alpha_1} x_2^{\alpha_2} \ldots x_n^{\alpha_n},$$

$$\alpha_1, \ldots, \alpha_n \in \mathbb{N}, \quad a_{\alpha_1\alpha_2\ldots\alpha_n} \in K.$$

A polynomial is said to be **homogeneous of degree** d if all the terms which effectively appear in it (that is appear with non-zero coefficient) are of degree d, i.e. satisfy $\alpha_1+\alpha_2+\ldots+\alpha_n = d$.

Examples. $3x^3+2x^2-x+1$ is a polynomial of *degree* 3 (i.e., a *cubic*), x^3+2x^2-x+2 is a *monic* polynomial. $2x^2y+3xy-2$ is a polynomial in the two *indeterminates* x and y. $2x^2y+3xy^2+y^3$ is a *homogeneous* polynomial of *degree* 3.

Given a polynomial $f \in K[x]$ one defines a function $S_f: K \rightarrow K$, '*substituting* in f', defined by

$$S_f(c) = f_0+f_1c+f_2c^2+\ldots+f_Nc^N.$$

S_f is known as a **polynomial function.** $S_f(c)$ is customarily denoted by $f(c)$ and is known as the value of f at c.

Note. In these definitions a *polynomial form* f, i.e., the sequence of coefficients occurring in our 'naive' polynomial, is an element of $K[x]$, whereas the corresponding *polynomial function,* i.e. the function obtained by 'evaluating' the polynomial, is a mapping $K \rightarrow K$. Thus $f \in K^{\mathbb{N}}$, $S_f \in K^K$.

If $f(c) = 0$, we say that c is a **zero** or **root** of f in K.

Traditionally c was called a *root* of the polynomial equation $f(x) = 0$ or a *zero* of the polynomial $f(x)$. Nowadays, root and zero tend to be used as synonyms, possibly so as to avoid expressions such as 'a zero of a non-zero polynomial'.

It can be shown that if D is an integral domain and c is a zero of the polynomial $f \in D[x]$ in D, then

$$f = q.r$$

where $q \in D[x]$ and r is the linear polynomial $x-c$, i.e.

$$r = (-c, 1, 0, 0, \ldots).$$

If $f = q.r^k$ and $q(c) \neq 0$, then we say that c is a zero of f with **multiplicity** k. A zero having multiplicity 1 is said to be **simple.** A polynomial all of whose zeros are simple is said to have **distinct** zeros.

A polynomial $f \in D[x]$ of degree N can be shown to have at most N zeros in D.

If $f = f_0+f_1x+f_2x^2+\ldots+f_Nx^N \in K[x]$, then we define the **derivative** of f to be the polynomial

$$Df = f' = f_1+2f_2x+3f_3x^2+\ldots+Nf_Nx^{N-1}.$$

Note. (i) This definition is a purely algebraic one as is the operation D which it describes.

(ii) In general, if K is a commutative ring and $D: K \to K$ is such that

$$D(f+g) = Df+Dg,$$

$$D(fg) = (Df)g+f(Dg) \quad \text{for all} \quad f, g \in K,$$

then we say that D is a **derivation** of K.

Example. Let f be the polynomial $x^3-x^2-x+1 \in \mathbb{Z}[x]$. Then $S_f(c)$ is the function defined by $S_f(c)$ $(=f(c)) = c^3-c^2-c+1$. We note that

$$f(1) = f(-1) = 0,$$

i.e. both 1 and -1 are *zeros* of f in $\mathbb{Z}$. $f = (x^2-2x+1)(x+1) = qr$, say, where q is the polynomial x^2-2x+1 and r the *linear polynomial* $x+1$. Since $q(-1) \neq 0$ it follows that -1 is a *simple* zero of f. However, it can be checked that 1 is a zero of *multiplicity* 2. The *derivative* of f, Df, is the polynomial $3x^2-2x-1$.

12 Groups II

Let ϕ, ψ be two automorphisms (p. 34) of a group G. If we define the product of ϕ and ψ, $\phi\psi : G \to G$, by $\phi\psi(g) = \phi(\psi(g))$, then under this operation the set of all automorphisms of G forms a group – the **group of automorphisms** of G, denoted by $\mathrm{Aut}(G)$ – a subgroup of the group of permutations on the *set* G.

In particular, given a fixed element $a \in G$ we can define an automorphism of G, **conjugation** by a, by

$$g \mapsto a^{-1}ga \quad \text{all} \quad g \in G.$$

An automorphism of this kind is known as an **inner automorphism.** An element g and its image under an inner automorphism are known as **conjugate elements.**

The set of all inner automorphisms forms a normal subgroup of $\mathrm{Aut}(G)$ which is denoted by $\mathrm{In}(G)$. Elements of the set $\mathrm{Aut}(G) - \mathrm{In}(G)$ (or, alternatively, but not equivalently, of the group $\mathrm{Aut}(G)/\mathrm{In}(G)$) are called **outer automorphisms** of G.

If $g \mapsto a^{-1}ga$ is the identity automorphism $g \mapsto g$, i.e. if $ag = ga$ for all $g \in G$, then a is called a **central** element of G. The set $Z(G)$ (or Z where no confusion is likely to arise) of all central elements of G is called the **centre** of G. (Z is a normal subgroup of G.)

A group is said to be **complete** if it has no outer automorphisms, i.e. if $\mathrm{In}(G) = \mathrm{Aut}(G)$, and $Z(G) = \{e\}$.

If M is a subset of G, then the set $C_G(M)$ of elements of G which commute with every element of M, i.e. the set

$$\{c \in G \mid cm = mc \quad \text{for all} \quad m \in M\},$$

is called the **centraliser** of M. ($C_G(M)$ is a subgroup of G.) The set

$$N_G(M) = \{n \in G \mid n^{-1}mn \in M \text{ for all } m \in M\}$$

is called the **normaliser** of M in G. (If M is a subgroup of G, then so is its normaliser.) The set $D_m = \{\phi(m) \mid \phi \in \mathrm{In}(G)\}$, i.e., the set of all elements of G which are conjugate to m, is known as the **conjugate class** of m. G is partitioned by conjugate classes of which any two are equal or disjoint. If Δ is the set of all conjugate classes, and if g_D

denotes a representative of the conjugate class D, then the **class equation** states that, for a finite group G,

$$[G:\{e\}] = \sum_{D\in\Delta} [G:C_G(g_D)], \quad g_D \in D.$$

A link between many of these definitions is provided by the notion of a group operating on a set. We say that a group G **operates on** or **acts on** a set X if we are given a mapping $G\times X \to X$, denoted by $(g, x) \mapsto g.x$, which satisfies:

$$(a)\ g.(h.x) = (gh).x, \qquad (b)\ e.x = x,$$

for all $g, h \in G$ and $x \in X$.

If G operates on X, then for each $x \in X$, the elements $g \in G$ such that $g.x = x$ form a subgroup of G known as the **stabiliser** of x in G. The set of elements of the form $g.x$ for fixed $x \in X$ and varying g in G is called the **orbit** of x **under** G and is denoted by $G.x$.

G is said to operate **transitively** on X when to each pair of points $x, y \in X$ there is at least one $g \in G$ such that $g.x = y$, i.e. when the orbit of every element $x \in X$ is the set X itself.

Note. (i) It must be noted that a group G can operate on a set X in different ways and that the different operations will yield different stabilisers and orbits (see below).

(ii) In particular, G can be made to operate on itself or on a subset S of itself, or a subgroup S can be made to operate on G, in several ways, for example,

$$(a)\quad (g, h) \mapsto gh,$$

or

$$(b)\quad (g, h) \mapsto g^{-1}hg.$$

We note that in case (a) the *orbit* of h when S, a subgroup of G, operates on G, is the *right coset* Sh (p. 28). In example (b) the *orbit* of h when G acts on itself will be the *conjugate class* of h, and the *stabiliser* of h in G will be the *centraliser* (and *normaliser*) of h in G, i.e. the subgroup $C_G(h)$.

In general, two orbits Gx_1 and Gx_2 are either disjoint or equal and X is partitioned (p. 19) by orbits (and thus, say, a group G is partitioned by cosets or by conjugate classes).

Example. Let r and n be positive integers satisfying $r \leqslant n$ and let $\mathscr{P}r(A)$ be the set of all subsets of order r of the set $A = \{1, 2, \ldots, n\}$. S_n, the symmetric group (p. 27) of degree n, *acts* on $\mathscr{P}r(A)$ if we define $\sigma.U$, where $\sigma \in S_n$ and $U \in \mathscr{P}r(A)$, to be the set $\{\sigma(u) \mid u \in U\}$. S_n acts *transitively* on $\mathscr{P}r(A)$, i.e. the *orbit* of any element of $\mathscr{P}r(A)$ is $\mathscr{P}r(A)$. The *stabiliser* of U is the set of all permutations which interchange the elements of U. There are $r!$ ways in which the elements of U can be rearranged and corresponding to each of these there will be $(n-r)!$ elements of S_n which rearrange the remaining $n-r$ elements of $A-U$. Hence,the *stabiliser* of U contains $r!(n-r)!$ elements.

The cardinal of the *orbit* of U (i.e., the cardinal of $\mathscr{P}r(A)$ and, hence, the number of ways in which one can select r objects from n) will therefore be $\#(S_n)/(r!(n-r)!) = n!/(r!(n-r)!)$.

A subgroup H of G which is mapped *onto* itself by all automorphisms of G is called a **characteristic subgroup** of G. Subgroups which are mapped *into* themselves by all endomorphisms of G are said to be **fully invariant.**

Given two elements $a, b \in G$, we define the **commutator** of a and b, denoted by $[a, b]$, to be the element $a^{-1}b^{-1}ab$. Given two subsets A, B of G, we define the **commutator group** of these subsets, denoted by $[A, B]$, to be the subgroup of G generated by all commutators of the form $[a, b]$ where $a \in A$ and $b \in B$. In particular, the **derived** or **commutator subgroup** of G is the subgroup $[G, G]$ of G generated by the set of commutators of every pair of elements of G.

The notion of *centre* and *derived* group can be extended in the following way:

let $Z_0(G)$ denote $\{e\}$,

$Z_1(G)$ denote the *centre* of G,

and define $Z_{i+1}(G)$, $i = 1, 2, \ldots$,

to be the subgroup of G satisfying

$$Z_{i+1}(G)/Z_i(G) = Z(G/Z_i(G)).$$

This defines a sequence of subgroups of G satisfying

$$\{e\} = Z_0(G) \lhd Z_1(G) \lhd Z_2(G) \lhd \ldots \lhd Z_{k+1}(G) \lhd \ldots.$$

This chain (p. 75) is known as the **upper central series** of G. If, for some p, $Z_p(G) = G$, then we say that G is a **hypercentral** group (or *nilpotent*, see note on p. 61).

Similarly, we define

$$G^{(0)} = G,$$

$$G^{(1)} = G' = [G, G],$$

$$G^{(k+1)} = (G^{(k)})' = [G^{(k)}, G^{(k)}] \quad (k = 1, 2, 3, \ldots),$$

and obtain the **derived series** or **commutator series**

$$G = G^{(0)} \rhd G^{(1)} \rhd G^{(2)} \rhd \ldots \rhd G^{(k+1)} \rhd \ldots.$$

If, for some q, $G^{(q)} = \{e\}$, then we say that G is **solvable.**

An alternative, equivalent definition is that G is *solvable* if and only if there exists a series of subgroups

$$G = S_0 \supset S_1 \supset S_2 \supset \ldots \supset S_n = \{e\},$$

such that for all $k = 1, \ldots, n$,

(*a*) S_k is normal in S_{k-1},

(*b*) S_{k-1}/S_k is Abelian.

The **lower central series** of a group G is the chain of subgroups defined recursively by

$$G = G_1, \quad G_{k+1} = [G_k, G], \quad k = 1, 2, 3, \ldots.$$

A group G for which $G_N = \{e\}$ for some positive integer N, and for which $G_{N-1} \neq \{e\}$, is said to be **nilpotent of class** $N-1$.

Note. (i) G is **nilpotent** if and only if it is **hypercentral.**
(ii) Any nilpotent group is solvable. S_3 (p. 27) is solvable but not nilpotent.

In general, a series

$$G = N_0 \supset N_1 \supset N_2 \supset \ldots \supset N_k = \{e\}$$

is called a **subinvariant** (or **subnormal**) **series** when $N_i \lhd N_{i-1}$ for all $i = 1, \ldots, k$. If, further, $N_i \lhd G$, $i = 1, \ldots, k$, then the series is said to be a **normal series.**

G is **simple** when it admits no normal series with $k > 1$, i.e. when it has no proper normal subgroups. A **refinement** of a series S is any series obtained from S by the insertion of additional subgroups in the chain. A **composition series** is a subinvariant series for which N_{i-1}/N_i is simple and non-trivial, $i = 1, \ldots, k$, i.e. it is a subinvariant series having no proper refinements. The quotient groups N_{i-1}/N_i are known as the **factors** of the subinvariant series.

Example. The *Cayley table* of the group D_4 (see p. 27) is shown below. Here M_4 at the intersection of the row labelled M_3 and the column labelled R_2 indicates that the result of compounding the maps (first) R_2 and (then) M_3 is the transformation M_4 ($= M_3R_2$).

	I	R_1	R_2	R_3	M_1	M_2	M_3	M_4
I	I	R_1	R_2	R_3	M_1	M_2	M_3	M_4
R_1	R_1	R_2	R_3	I	M_3	M_4	M_2	M_1
R_2	R_2	R_3	I	R_1	M_2	M_1	M_4	M_3
R_3	R_3	I	R_1	R_2	M_4	M_3	M_1	M_2
M_1	M_1	M_4	M_2	M_3	I	R_2	R_3	R_1
M_2	M_2	M_3	M_1	M_4	R_2	I	R_1	R_3
M_3	M_3	M_1	M_4	M_2	R_1	R_3	I	R_2
M_4	M_4	M_2	M_3	M_1	R_3	R_1	R_2	I

The *centre* of D_4 is $\{I, R_2\}$. A typical *inner automorphism* of D_4 is given by *conjugation* by M_1, i.e. $I \mapsto I$, $R_1 \mapsto R_3$, $R_2 \mapsto R_2$, $R_3 \mapsto R_1$, $M_1 \mapsto M_1$, $M_2 \mapsto M_2$, $M_3 \mapsto M_4$, $M_4 \mapsto M_3$. The *centraliser* of M_1 is the subgroup $\{I, R_2, M_1, M_2\}$.

The *normaliser* of $\{I, R_1\}$ is $\{I, R_1, R_2, R_3\}$. The *conjugate classes* of D_4 are $\{I\}$, $\{R_2\}$, $\{R_1, R_3\}$, $\{M_1, M_2\}$, $\{M_3, M_4\}$. The *centralisers* of class representatives are D_4, D_4, $\{I, R_1, R_2, R_3\}$, $\{I, R_2, M_1, M_2\}$ and $\{I, R_2, M_3, M_4\}$ and the *class equation* confirms that $8 = 1+1+2+2+2$. The *centre* of D_4 is both *characteristic* and *fully invariant* (the former is true of the centre of any group, the latter is not).

The *commutator* of R_1 and M_1, $[R_1, M_1]$, is $R_1^{-1}M_1^{-1}R_1M_1 = M_4M_3 = R_2$. The *derived group*, $G^{(1)} = [G, G]$, is the subgroup $\{I, R_2\}$.

The factor group $D_4/Z_1(D_4)$ is isomorphic to Klein's four group (p. 207) which is Abelian and hence its own centre. Hence $Z(D_4/Z_1(D_4))$ has order 4 and if $Z_2(D_4)/Z_1(D_4)$ is to have order 4, then we must have $Z_2(D_4) = D_4$, i.e., the *upper central series* for D_4 is $\{I\} \lhd Z(D_4) \lhd D_4$, and D_4 is *hypercentral*. The *derived series* for D_4 is similarly found to be $D_4 \rhd G^{(1)} \rhd \{I\}$. D_4 is therefore, *solvable*. The *lower central series* for D_4 is $D_4 = G_1 \rhd [G_1, G_1] \rhd \{I\}$, i.e. D_4 is *nilpotent* of *class* 2. A *composition series* for D_4 is

$$D_4 \rhd \{I, R_1, R_2, R_3\} \rhd \{I, R_2\} \rhd \{I\}.$$

All the *factors* of this series are isomorphic to the cyclic group of order 2.

Free groups, generators and relations

Let $M = \{a, b, c, \ldots\}$ be a set of distinct symbols and let $a^{-1}, b^{-1}, c^{-1}, \ldots$ be a further set of symbols M', disjoint from M and in one–one correspondence with it.

We define a **word** in $N = M \cup M'$ to be a finite sequence $f_1 f_2 \ldots f_n$ where $f_i \in N$, $i = 1, \ldots, n$, and we say that the word is of **length** n. The set of all words, S, together with the operation of juxtaposition is a semigroup (p. 25) and is known as the **free semigroup** on N. We introduce the empty word, 1, as the word of zero length, thus making S a monoid (p. 25), and define an equivalence relation on S by sRs' if and only if s' can be *derived* from s by the insertion or deletion of a finite number of words of the type aa^{-1}, $a^{-1}a$, bb^{-1}, $b^{-1}b$, ... (In particular, $aa^{-1}R1$.) If we now define the inverse of the equivalence class $R(f_1 f_2 \ldots f_n)$ to be the class $R(f_n^{-1} f_{n-1}^{-1} \ldots f_1^{-1})$, with the convention that $(a^{-1})^{-1} = a$, etc., then S/R is a group. It is known as the **free group** on M. M is said to be a **free set of generators** of this group and the cardinality of M is called the **rank** of the group.

Any group may be derived from a suitable free semigroup S by a suitable choice of equivalence relation R. In general we take M to be a set of symbols, $a, b, c, \ldots$, in one–one correspondence with a set of generators G and define R so that the equivalence class determined by the empty word contains all words which we wish to associate with the identity element of G. Such words are called **relators** and an expression $W_1 = W_2$, where W_1 and W_2 are words and $W_1W_2^{-1}$ is a relator, is called a **relation** (W_1 and W_2 then represent the same element of G).

If every relator can be derived from a given set of relators, $P_1, P_2, P_3, \ldots$, by means of suitable insertions and deletions, we say that $P_1, P_2, P_3, \ldots$ is a **set of defining relators** for the group G on the generators $a, b, c, \ldots$, and write

$$G = \langle a, b, c, \ldots; P_1, P_2, P_3, \ldots\rangle.$$

G is then said to be **presented** in terms of generators and relators.

Note. (i) G can also be presented in terms of generators and relations, e.g., the quaternion group, a group of order 8, can be presented in terms of *relators* as $\langle a, b; a^4, a^2b^{-2}, b^{-1}aba\rangle$ or in terms of *relations* as

$$\langle a, b; a^4 = I, a^2 = b^2, b^{-1}ab = a^{-1}\rangle.$$

Presentation by means of relations is the more usual.

(ii) When presenting a group it is the convention to omit mention of trivial relators of the type aa^{-1}, $a^{-1}a$. A free group of rank 2 will therefore have a presentation $\langle a, b\rangle$; being free it can be presented in terms of generators alone – there being, in this presentation, no non-trivial relators.

If the set $\{a, b, c, \ldots\}$ is finite, then G is said to be **finitely generated.** If, in addition, the set of defining relators (or relations) is finite, then G is said to be **finitely presented.**

The problem of determining whether, when given a group in terms of generators and relations, one can decide in a finite number of steps whether or not an arbitrary word W defines the identity element of G is known as the **word problem.**

Given two groups, G and H, such that $G \cap H = \{e\}$,

$$G = \langle g_1, g_2, \ldots; P_1, P_2, \ldots\rangle$$

and

$$H = \langle h_1, h_2, \ldots; Q_1, Q_2, \ldots\rangle,$$

we define the **direct product**, $G \times H$, of G and H to be

$$\langle g_1, g_2, \ldots, h_1, h_2, \ldots; P_1, P_2, \ldots, Q_1, Q_2, \ldots, h_ig_jh_i^{-1}g_j^{-1} \text{ all } i, j\rangle,$$

and the **free product**, $G * H$, of G and H to be

$$\langle g_1, g_2, \ldots, h_1, h_2, \ldots; P_1, P_2, \ldots, Q_1, Q_2, \ldots\rangle.$$

G and H are known as the **free factors** of $G * H$. If a group cannot be expressed as the free product of two factors, neither of which is non-trivial, then we say that it is **indecomposable** with respect to free products.

An alternative, and more straightforward, definition of the *direct product* is that $G \times H$ consists of all ordered pairs (g, h) of elements, $g \in G$, $h \in H$, with the product of pairs defined by

$$(g_1, h_1)(g_2, h_2) = (g_1 g_2, h_1 h_2).$$

When considering Abelian groups it is more usual to adopt the additive notation for the group operation. In this case we define the **direct sum**, $G \oplus H$, of G and H in an analogous manner to the direct product, i.e. as the set of all ordered pairs (g, h) with the sum of pairs defined by

$$(g_1, h_1) + (g_2, h_2) = (g_1 + g_2, h_1 + h_2).$$

Example. In addition to groups isomorphic to the quaternion group (p. 63) and the group D_4 (p. 61), there are three other types of group of order 8, namely, those isomorphic to the cyclic group $C_8 = \langle a; a^8 = I \rangle$ to the *direct sum*

$$C_4 \oplus C_2 = \langle a; a^4 = I \rangle \oplus \langle b; b^2 = I \rangle = \langle a, b; a^4 = b^2 = I, ab = ba \rangle,$$

and to the *direct sum*

$$C_2 \oplus C_2 \oplus C_2 = \langle a, b, c; a^2 = b^2 = c^2 = I, ab = ba, bc = cb, ca = ac \rangle.$$

See also: Fundamental Theorem of Abelian Groups (p. 198); Burnside's problem (p. 198); The Isomorphism Theorems (p. 205); Jordan–Hölder Theorem (p. 207); Klein's four group (p. 207).

13 Number systems II

Real numbers

In § 4 it was shown how, beginning with $\mathbb{N}$, it is possible to construct $\mathbb{Z}$ and $\mathbb{Q}$ by considering quotient sets of suitable Cartesian products. It is not possible to construct $\mathbb{R}$, the set of all real numbers, in a similar fashion and other methods must be employed. One, due essentially to Dedekind, is given below; an alternative approach is given on p. 113.

We first note that the order relation $<$ defined on $\mathbb{Q}$ (p. 24) has the property that, given $a, b \in \mathbb{Q}$ such that $a < b$, there exists $c \in \mathbb{Q}$ such that $a < c$ and $c < b$. (Note that $(\mathbb{Z}, <)$, for example, does not possess this property.)

Now consider ordered pairs of elements of $\mathscr{P}(\mathbb{Q})$ (p. 9), (A, B) say, satisfying:

(i) $A \cup B = \mathbb{Q}, A \cap B = \varnothing$,
(ii) A and B are both non-empty,
(iii) $a \in A$ and $b \in B$ together imply $a < b$.

Such a pair of sets (A, B) is known as a **Dedekind cut** or **section.**

An equivalence relation R is defined upon the set of cuts by $(A, B)\, R\, (C, D)$ if and only if there is at most one rational number which is either in both A and D or in both B and C.

This ensures that the cuts $(\{x|x \leqslant q\}, \{x|x > q\})$ and $(\{x|x < q\}, \{x|x \geqslant q\})$ are equivalent for all $q \in \mathbb{Q}$.

Each equivalence class under this relation is defined to be a **real number.** The **set of all real numbers,** denoted by $\mathbb{R}$, is then the set of all such equivalence classes.

The class containing the cut $(\{x|x \leqslant 0\}, \{x|x > 0\})$ is known as the **real number zero.**

If the class contains a cut (A, B) such that A contains positive rationals, then the class is a **positive real number,** if B should contain negative rationals then the class is a **negative real number.**

Thus, for example, '$\sqrt{2}$' which contains the cut $(\{x|x^2 < 2\}, \{x|x^2 > 2\})$ is positive since $1 \in A = \{x|x^2 < 2\}$.

To define **addition** of real numbers we must consider cuts (A_1, B_1) and (A_2, B_2) representing the real numbers α_1 and α_2. We define $\alpha_1+\alpha_2$ to be the class containing the cut (A_3, B_3) where A_3 consists of all the sums $a = a_1+a_2$ obtained by selecting a_1 from A_1 and a_2 from A_2.

Given the real number α, represented by the cut (A_1, B_1), we define $-\alpha$, **negative** α, to be the class containing the cut $(-B_1, -A_1)$ defined by $a \in A_1 \Leftrightarrow -a \in -A_1$ and $b \in B_1 \Leftrightarrow -b \in -B_1$. It will be observed that $\alpha+(-\alpha) = 0$, and that **subtraction** can now be defined by $\alpha-\beta = \alpha+(-\beta)$.

Of two non-zero numbers α and $-\alpha$, one is always positive. The one which is positive is known as the **absolute value** or **modulus** of α and is denoted by $|\alpha|$. Thus $|\alpha| = \alpha$ if α is positive and $|\alpha| = -\alpha$ if α is negative. $|0|$ is defined to be 0.

If α_1 and α_2 are two *positive* real numbers, then the **product** $\alpha_1\alpha_2$ is the class containing the cut (A_4, B_4) where A_4 consists of the negative rationals, zero, and all the products $a = a_1 a_2$ obtained by selecting a positive a_1 from A_1 and a positive a_2 from A_2.

The definition is extended to negative numbers by agreeing that if α_1 and α_2 are positive, then

$$(-\alpha_1)\alpha_2 = \alpha_1(-\alpha_2) = -\alpha_1\alpha_2, \quad (-\alpha_1)(-\alpha_2) = \alpha_1\alpha_2.$$

Finally, we define

$$0\alpha = \alpha 0 = 0 \quad \text{for all } \alpha.$$

With these definitions it can be shown that the real numbers $\mathbb{R}$ form an *ordered field*.

By associating the element $q \in \mathbb{Q}$ with the class containing the cut $(\{x|x \leqslant q\}, \{x|x > q\})$, one can define a monomorphism (of fields) (p. 34) $\mathbb{Q} \to \mathbb{R}$. We can, therefore, consider $\mathbb{Q}$ to be a subfield of $\mathbb{R}$ (i.e. identify $\mathbb{Q}$ with a subfield of $\mathbb{R}$). Those elements of $\mathbb{R}$ which do not then belong to $\mathbb{Q}$ are known as **irrational numbers**.

An important property of $\mathbb{R}$ which can now be established is that given any non-empty subset $V \subset \mathbb{R}$ for which there exists an **upper bound,** M, i.e. an element $M \in \mathbb{R}$ such that $v \leqslant M$ for all $v \in V$, then there exists a **least upper bound (lub)** or **supremum (sup)** L such that if M is any upper bound of V, then $L \leqslant M$.

An ordered field F in which every non-empty subset V possessing an upper bound has a least upper bound is said to satisfy the **lub principle.** It can be shown that an ordered field satisfies the lub principle if and only if it is complete (p. 109).

In a similar manner, we can define a **greatest lower bound (glb)** or **infimum (inf)** for any non-empty subset V of $\mathbb{R}$ which possesses a lower bound.

It will be observed that not every set in $\mathbb{R}$ possesses an upper (or lower) bound in $\mathbb{R}$, for example, $\mathbb{N}$. In order to overcome certain consequences of this, one often makes use in analysis of the **extended real number system**, $\overline{\mathbb{R}}$, consisting of $\mathbb{R}$ together with the two symbols $-\infty$ and $+\infty$ having the properties:

(*a*) If $x \in \mathbb{R}$, then

$$-\infty < x < +\infty,$$

and
$$x+\infty = +\infty, \quad x-\infty = -\infty, \quad \frac{x}{+\infty} = \frac{x}{-\infty} = 0.$$

(*b*) If $x > 0$, then

$$x.(+\infty) = +\infty, \quad x.(-\infty) = -\infty.$$

(*c*) If $x < 0$, then

$$x.(+\infty) = -\infty, \quad x.(-\infty) = +\infty.$$

Note. (i) $\overline{\mathbb{R}}$ does not possess all the algebraic properties of $\mathbb{R}$.

(ii) If a non-empty set V does not possess an upper (lower) bound in $\mathbb{R}$, we say $\text{lub}\, V = +\infty$ ($\text{glb}\, V = -\infty$).

Complex numbers

The field of **complex numbers**, $\mathbb{C}$, can now be defined in several alternative ways.

Consider $\mathbb{R}\times\mathbb{R}$ and take the elements of $\mathbb{C}$ to be ordered pairs $(x, y) \in \mathbb{R}\times\mathbb{R}$. The operations of **addition** and **multiplication** of elements of $\mathbb{C}$, i.e. of complex numbers, are defined by

$$(x_1, y_1)+(x_2, y_2) = (x_1+x_2, y_1+y_2),$$

and
$$(x_1, y_1)\times(x_2, y_2) = (x_1x_2-y_1y_2, x_1y_2+x_2y_1),$$

and routine calculations suffice to show that $\mathbb{C}$ is then a field.

In an analogous manner to that followed on previous occasions, we can define a monomorphism (of fields) (p. 34),

$$r: \mathbb{R} \to \mathbb{C}, \quad \text{by} \quad r(x) = (x, 0).$$

This enables us to regard $\mathbb{R}$ as a subfield of $\mathbb{C}$.

It can, moreover, be easily checked that

$$(x, y) = (x, 0) + (0, 1)\times(y, 0)$$

and so, making use of the monomorphism defined above, one can write

$$(x, y) = x+\mathrm{i}y$$

where $x, y \in \mathbb{R}$ and $\mathrm{i} = (0, 1)$.

Note. (i) It is seen that $\mathrm{i}^2 = (0, 1) \times (0, 1) = (-1, 0) = -1$.
(ii) Engineers, physicists and others frequently denote i $(=\sqrt{-1})$ by j.

Given a complex number $z = x+\mathrm{i}y$, where $x, y \in \mathbb{R}$, we say that x is the **real part,** $\mathscr{R}(z)$, and y the **imaginary part,** $\mathscr{I}(z)$, of z. The number $x-\mathrm{i}y$ is known as the **complex conjugate** of z and is denoted by $\bar{z}$ (or z^*).

The construction outlined above for $\mathbb{C}$ suggests an obvious geometrical representation of the complex numbers as points of the Cartesian plane $\mathbb{R}\times\mathbb{R}$. This representation is known as an **Argand diagram.**

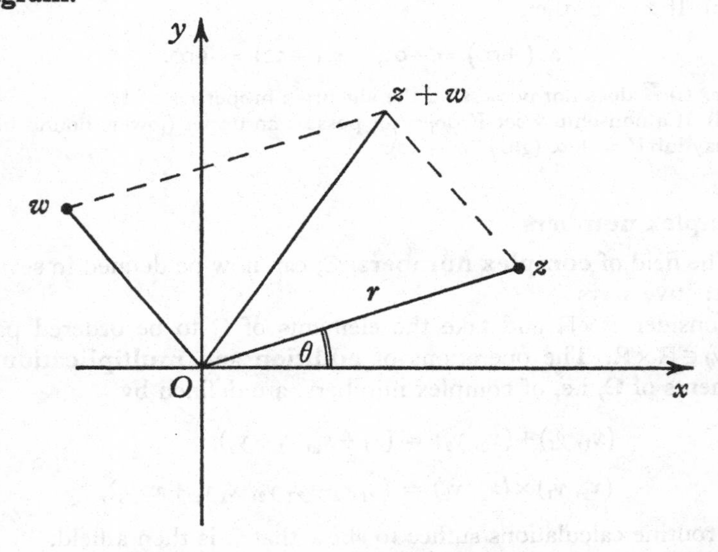

The diagram is based on a pair of perpendicular coordinate axes in the plane. The number $z = x+\mathrm{i}y$ is associated with the point with coordinates (x, y). With this representation, the addition of complex numbers is interpreted as the addition of vectors in the plane (see p. 169). The 'length', r, of the segment Oz, $\sqrt{(x^2+y^2)}$, is known as the **absolute value** or **modulus** of z and is denoted by $|z|$. We therefore have

$$|z| = r = \sqrt{(x^2+y^2)} = \sqrt{z\bar{z}}.$$

The angle which the segment Oz makes with the Ox axis is known as the **argument (amplitude** or **angle)** of z and is denoted by $\arg z$.

We therefore have

$$\tan(\arg z) = \tan\theta = y/x,$$

and $\arg z$ is defined as a real number modulo 2π (provided $z \neq 0$, for $\arg 0$ is not defined).

Note. Some authors take $0 \leqslant \arg z < 2\pi$, while others opt for $-\pi < \arg z \leqslant \pi$. This restricted value of the argument is often known as the **principal argument.**

Example. Let $z_1 = 1+i$, $z_2 = 1-i\sqrt{3}$. Then $\bar{z}_1 = 1-i$, $\bar{z}_2 = 1+i\sqrt{3}$, $|z_1| = \sqrt{2}$, $\arg z_1 = \frac{1}{4}\pi$, $|z_2| = 2$, $\arg z_2 = \frac{5}{3}\pi$, $z_1 z_2 = (1+\sqrt{3})+(1-\sqrt{3})i$, $|z_1 z_2| = 2\sqrt{2}$, $\arg(z_1 z_2) = \frac{23}{12}\pi$.

The coordinates r and θ are known as **polar coordinates** (see note (iii) below). The connections between the polar coordinates of a point and the complex number $z = x+iy$ which it represents are

$$x = r\cos\theta, \quad y = r\sin\theta,$$

$$r = |z| = \sqrt{(x^2+y^2)},$$

$$\theta = \arg z,$$

$$z = r(\cos\theta + i\sin\theta).$$

Note. (i) The geometry of the triangle suggests the inequality

$$|z_1+z_2| \leqslant |z_1|+|z_2|,$$

known, for that reason, as the **triangle inequality.** The two formulae

$$|z_1 z_2| = |z_1|\,|z_2| \quad \text{and} \quad \arg(z_1 z_2) \equiv \arg(z_1)+\arg(z_2) \pmod{2\pi}$$

allow one to give a geometrical interpretation of the multiplication of complex numbers.

(ii) Alternative constructions of $\mathbb{C}$ are:

(*a*) Consider the set, M, of all matrices of $M_2(\mathbb{R})$ of the form

$$\begin{pmatrix} a & -b \\ b & a \end{pmatrix}.$$

It is easy to check that under matrix addition and multiplication M forms a field with zero O_2 and identity element I_2.

Moreover, $p : \mathbb{R} \to M$ defined by

$$p(c) = \begin{pmatrix} c & 0 \\ 0 & c \end{pmatrix}$$

is a field isomorphism between $\mathbb{R}$ and a subfield of M.

Mapping the matrix $\begin{pmatrix} a & -b \\ b & a \end{pmatrix}$

onto $a+ib$, we obtain a field isomorphism between M and $\mathbb{C}$ as previously defined.

(*b*) We define $\mathbb{C}$ to be the quotient ring (p. 36) of $\mathbb{R}[t]$ by the principal ideal (p. 32) (t^2+1), i.e. $\mathbb{C} = \mathbb{R}[t]/(t^2+1)$. If we denote the image of the polynomial $t \in \mathbb{R}[t]$ under the canonical mapping by i, then every element of $\mathbb{C}$ can be written uniquely as $x+\mathrm{i}y$ where $x, y \in \mathbb{R}$. The operations of addition and multiplication on $\mathbb{C}$ are the natural operations of the quotient algebra.

(iii) So as to facilitate the study of certain curves (e.g. the equiangular spiral) one frequently relaxes the conditions $r \geqslant 0$ and $0 \leqslant \theta < 2\pi$ on *polar coordinates*. One then has an extended system of **polar coordinates** in which r and θ can take all real values. In the extended system any pair (ρ, w) will determine a unique point of the plane, yet every point in the plane will possess an infinite number of polar coordinates, namely $(\rho, w+2n\pi)$, $(-\rho, w+(2n+1)\pi)$ for all $n \in \mathbb{Z}$. (In the extended system the equiangular spiral is described by the single equation $r = \mathrm{e}^{a\theta}$, whereas in the restricted system a whole set of equations would be required to define it.)

The quaternion algebra

We have shown how, by defining multiplication suitably on $\mathbb{R} \times \mathbb{R}$, it is possible to construct a field $\mathbb{C}$ which is an extension (p. 72) of $\mathbb{R}$. Indeed, since $\mathbb{C}$ is a vector space of dimension 2 over $\mathbb{R}$, $\mathbb{C}$ is a *commutative algebra* (p. 39) over $\mathbb{R}$. It is natural to attempt to repeat this process and to try to embed $\mathbb{C}$ in an algebra defined upon $\mathbb{R}^n$ $(n > 2)$. It is, in fact, impossible to find such an extension satisfying the field axioms, but, as the following construction shows, some measure of success can be attained.

Consider the vector space $\mathbb{R}^4$ generated by

$$1 = (1, 0, 0, 0),$$
$$\mathrm{i} = (0, 1, 0, 0),$$
$$\mathrm{j} = (0, 0, 1, 0),$$
$$\mathrm{k} = (0, 0, 0, 1),$$

and define multiplication of vectors in $\mathbb{R}^4$ by demanding that it should satisfy the ring axioms (with identity 1) and the rules $\mathrm{i}^2 = \mathrm{j}^2 = \mathrm{k}^2 = -1$, $\mathrm{ij} = \mathrm{k} = -\mathrm{ji}$, $\mathrm{jk} = \mathrm{i} = -\mathrm{kj}$, $\mathrm{ki} = \mathrm{j} = -\mathrm{ik}$. Multiplication, as thus defined, is clearly non-commutative, and so the resulting structure cannot be a field. It is a division ring (p. 31) and is known as the **ring of quaternions, Q**, or the **quaternion algebra.**

Q can also be obtained as the subring of $M_4(\mathbb{R})$ consisting of matrices of the form

$$\begin{pmatrix} x & -y & -z & -t \\ y & x & -t & z \\ z & t & x & -y \\ t & -z & y & x \end{pmatrix}.$$

See also: Dirichlet function (p. 202); Frobenius's Theorem (p. 203); Gaussian field (p. 203).

14 Fields and polynomials

Let F be a field. We say that a field K is an **extension** of F if F is a subfield of K.

If K is an extension of F, then K is a vector space over F and we define the **degree** of K over F, denoted by $[K:F]$, to be the dimension of K as a vector space over F. If $[K:F]$ is finite, K is said to be a **finite extension** of F or an **extension of F of finite degree.**

An element $x \in K$ is said to be **algebraic** over F if there exist elements $a_0, a_1, \ldots, a_n \in F$, not all of which are zero, such that

$$a_0 + a_1 x + a_2 x^2 + \ldots + a_n x^n = 0.$$

$x \in K$ is said to be **algebraic of degree n over** F if it is a root of a non-zero polynomial over F of degree n, but of no non-zero polynomial of lower degree.

When $x \in K$ is not algebraic over F, then it is said to be **transcendental** over F.

If, in particular, we take $F = \mathbb{Q}$ and $K = \mathbb{C}$, then those elements of $\mathbb{C}$ which are algebraic over $\mathbb{Q}$ are known as **algebraic numbers,** the remainder are **transcendental numbers.** An algebraic number satisfying an equation of the form

$$a_0 + a_1 x + a_2 x^2 + \ldots + a_{n-1} x^{n-1} + x^n = 0$$

where $a_0, a_1, \ldots, a_{n-1} \in \mathbb{Z}$, is said to be an **algebraic integer.**

If every element of K is algebraic over F, then we say that K is an **algebraic extension** of F.

A field F is said to be **algebraically closed** or **algebraically complete** if every polynomial with coefficients in F has a root in F.

If we define $d(f)$ to be the degree of the polynomial f, then $F[x]$, the ring of polynomials over F, is a Euclidean ring (p. 202) and so (see Unique Factorisation Theorem, p. 213) any polynomial in $F[x]$ can be written in a unique way as a product of irreducible polynomials which may have degree greater than one. It can be shown, however, that given any polynomial $f \in F[x]$ of degree $n > 1$, then there is an extension E of F (with $[E:F] \leqslant n!$) in which f has n roots and can be factored as a product of n linear factors. A field E is known as a **splitting field** (or **root field**) over F for f, if, in the ring of polynomials over E, but not over any subfield of E, f can be factored as a product of linear factors.

The extension K of F is a **simple extension** of F if it can be obtained by **adjoining** some single element $a \in K$ to F, i.e., if there exists $a \in K$ such that the smallest subfield of K containing F and a, denoted by $F(a)$ (or $F[a]$), is K itself.

A polynomial over F is **separable** if its irreducible factors do not have repeated roots. An element $a \in K$ is **separable** over F if it is a zero of a separable polynomial over F. An extension K over F is **separable** over F if all its elements are separable over F.

F is said to be **perfect** if all its finite extensions are separable.

Examples. $\mathbb{R}$ is an *extension* of the field $\mathbb{Q}$, and $\mathbb{C}$ an *extension* of $\mathbb{R}$. $\mathbb{C}$ is a *finite extension* of $\mathbb{R}$ and $[\mathbb{C}, \mathbb{R}] = 2$. $\mathbb{R}$ is not a finite extension of $\mathbb{Q}$. In $\mathbb{R}$, $\sqrt{2}$ is *algebraic of degree* 2 over $\mathbb{Q}$ since $\sqrt{2}$ is a root of the polynomial equation $x^2 - 2 = 0$ which has coefficients in $\mathbb{Q}$. (Since $\sqrt{2} \notin \mathbb{Q}$, $\mathbb{Q}$ is not *algebraically closed*.)

Since the polynomial $x^2 - 2$ is monic and has its coefficients in $\mathbb{Z}$, $\sqrt{2}$ is an *algebraic integer*. The set of all elements $a + b\sqrt{2}$, where $a, b \in \mathbb{Q}$, forms a field; it is a *simple extension* of $\mathbb{Q}$, and is denoted by $\mathbb{Q}(\sqrt{2})$. $\mathbb{Q}(\sqrt{2})$ is an *algebraic extension* of $\mathbb{Q}$ since $a + b\sqrt{2}$ is a root of $(x-a)^2 = 2b^2$, a polynomial with coefficients in $\mathbb{Q}$. Moreover, since a and $a + b\sqrt{2}$ are zeros of polynomials having no multiple roots, they are *separable* over $\mathbb{Q}$ and $\mathbb{Q}(\sqrt{2})$ is *separable* over $\mathbb{Q}$. $\mathbb{Q}$ (like all fields of characteristic 0) is *perfect*. $\mathbb{Q}(\sqrt{2})$ is a *splitting field* over $\mathbb{Q}$ for $x^2 - 2$.

$\mathbb{C} = \mathbb{R}(i)$ and is a *simple extension* of $\mathbb{R}$. It can be shown that a polynomial of $\mathbb{C}[z]$ is irreducible if and only if it is linear (see Fundamental Theorem of Algebra (p. 198)) and it follows that every polynomial of degree n in $\mathbb{C}[x]$ and, *a fortiori*, in $\mathbb{R}[x]$, has exactly n roots in $\mathbb{C}$. The non-real roots of a polynomial in $\mathbb{R}[x]$ will occur in *conjugate pairs* having the same multiplicity.

If G is the group of automorphisms (p. 34) of a field K and F is a subfield of K, then those elements of G which map every element of F onto itself, i.e., those $g \in G$ for which $g(a) = a$ for all $a \in F$, form a subgroup of G, known as the **group of automorphisms of** K **relative to** F (or **over** F) and denoted by $G(K, F)$. If the elements of F are the only elements of K left invariant under $G(K, F)$, i.e. if $g(a) = a$ for all $g \in G(K, F)$ implies $a \in F$, and if $[K : F]$ is finite then we say that K is a **normal extension** of F.

Equivalently, we say that a finite extension K of a field F is a *normal extension* of F if and only if K is the splitting field of a separable polynomial over F.

Given a polynomial $f \in F[x]$ with splitting field K over F, then we define the **Galois group** of f to be $G(K, F)$.

Example. Let $K = \mathbb{C}$ and $F = \mathbb{R}$. Then if σ is any *automorphism* of $\mathbb{C}$ *relative* to $\mathbb{R}$, $\sigma(\mathrm{i})^2 = \sigma(\mathrm{i}^2) = \sigma(-1) = -1$. Hence $\sigma(\mathrm{i}) = \pm\mathrm{i}$. From this it follows that $G(\mathbb{C}, \mathbb{R})$, the *Galois group* of, say, x^2+1, is of order 2, consisting of the identity automorphism ($\sigma_1(\mathrm{i}) = \mathrm{i}$) and the complex-conjugation automorphism ($\sigma_2(\mathrm{i}) = -\mathrm{i}$) (p. 68). Since σ_2 leaves only elements of $\mathbb{R}$ invariant, it follows that $\mathbb{C}$ is a *normal extension* of $\mathbb{R}$.

See also: Fundamental Theorem of Algebra (p. 198); Frobenius's Theorem (p. 203).

15 Lattices and Boolean algebra

On p. 18 a *poset* (partially ordered set), P, was defined to be a set together with a binary relation which was reflexive, antisymmetric and transitive. A poset is said to be a **chain** or **simply** (or **totally** or **linearly**) **ordered** when its elements satisfy

$$\forall a, b \in P, \quad a \leqslant b \quad \text{or} \quad b \leqslant a.$$

Here the binary relation is denoted by $\leqslant$. We shall see below that it is often convenient to use other notation for the relation, for example, $|$ and $\subset$.

A pair of elements of a general poset for which $a \leqslant b$ or $b \leqslant a$ are said to be **comparable.**

An element a of a poset P is said to be a **minimal element** of P if $x \leqslant a$, $x \in P$ together imply $x = a$. An element n of P satisfying $\forall x \in P$, $n \leqslant x$ is called a **null** (or **least**) **element** of P. (The null element is unique.) **Maximal** and **unit** (or **greatest**) **elements** of P are defined *dually* (that is, they are the minimal and null elements, respectively of the **dual poset** P^* defined by the *converse ordering relation* on the same elements, i.e., if $a \leqslant b$ in P, then $b \leqslant a$ in P^*).

The null and unit elements, when they exist, are usually denoted by O (or 0) and I (or 1).

We say that a **covers** b in the poset P if $a > b$ and if $a \geqslant x > b$ implies $x = a$.

Elements which cover the null element are called **atoms,** those which are covered by the unit element are called **coatoms (anti-atoms).**

Posets with only a finite number of elements can be exhibited by means of a **Hasse diagram.** A small circle is drawn to represent each element $a \in P$, a being placed higher than b whenever $a > b$. A straight segment is drawn from a to b whenever a covers b.

See example below and the diagram on p. 39 which is basically a Hasse diagram for the set {Abelian group, module, vector space, ring, algebra, commutative algebra} with the partial ordering induced by 'is weaker than'.

If Q is a subset of P and if there is an element $p \in P$ such that $q \leqslant p$ for all $q \in Q$, then p is called an **upper bound** of the set Q. Similarly, an element $l \in P$ such that $\forall q \in Q$, $l \leqslant q$, is called a **lower**

bound of Q. If the set of upper bounds has a least element, this element is called the **least upper bound, lub**, of Q (cf. p. 66); similarly, if the set of the lower bounds has a greatest element, this element is known as the **greatest lower bound, glb**, of Q. A poset in which lub Q and glb Q exist for all subsets Q is said to be **complete**.

A non-empty poset in which any two elements a and b have a glb (or **meet**), $a \wedge b$ (or $a \cap b$), and an lub (or **join**), $a \vee b$ (or $a \cup b$), is called a **lattice**.

It is a consequence of the definitions that on a lattice the two operations $\wedge$ and $\vee$ are **idempotent** (i.e. $a \wedge a = a = a \vee a$), *commutative*, *associative* and satisfy the **laws of absorption** (i.e. $x \wedge (x \vee y) = x \vee (x \wedge y) = x$).

A lattice, L, is **modular** if, whenever $a \leqslant c$, then

$$a \vee (b \wedge c) = (a \vee b) \wedge c.$$

A lattice in which the two distributive laws

$$(a) \quad a \vee (b \wedge c) = (a \vee b) \wedge (a \vee c)$$

and

$$(b) \quad a \wedge (b \vee c) = (a \wedge b) \vee (a \wedge c) \quad (\text{all } a, b, c \in L)$$

hold is called a **distributive** lattice.

Law (*a*) does, in fact, imply (*b*) and vice versa.

By a **complement** of an element a in a lattice L having null and unit elements we mean an element $b \in L$ such that

$$a \vee b = I, \quad a \wedge b = O.$$

The lattice is said to be **complemented** if all its elements have complements.

A complemented, distributive lattice is known as a **Boolean lattice** (or a **Boolean algebra**).

Example. The set {60, 30, 20, 15, 12, 10, 6, 5, 4, 3, 2, 1} together with the binary relation | defined by $a|b$ if a is a factor of b is a *poset* P. P is not a *chain* since 3 and 10 are not *comparable*. 1 is the *null* element of P and 60 is the *unit* element of P. P^*, the *dual* poset of P, has the same elements as P and the relation $\leqslant$ defined by $a \leqslant b$ if $b|a$. 60 is the *null* element of P^*. 20 *covers* 10 in the poset P. The *atoms* of P are 2, 3 and 5 and the *coatoms* are 12, 20 and 30.

The *Hasse diagram* for P is shown. If $Q = \{20, 15, 4\}$, then Q has the *least upper bound* 60 and *greatest lower bound* 1. Every two elements of P have a *meet* and a *join*, e.g. $12 \wedge 30 = 6$, $3 \vee 5 = 15$, and so P is a *lattice*. The complement of 3 is 20, for $3 \vee 20 = 60$ and $3 \wedge 20 = 1$. P is a *distributive* lattice but it is not *complemented*, since 2 has no *complement*, i.e. there is no

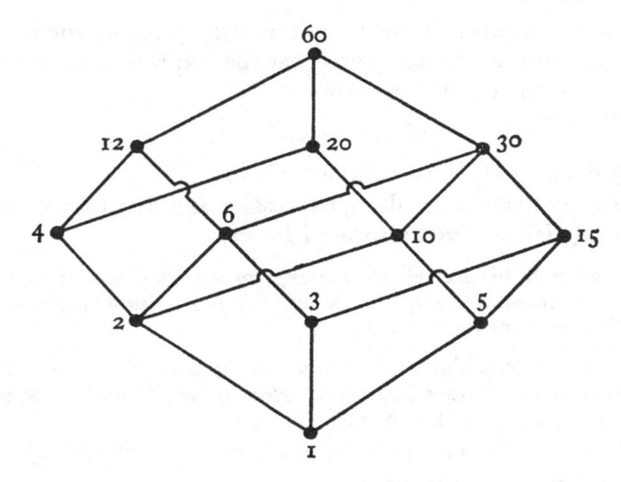

element a satisfying $a \vee 2 = 60$, $a \wedge 2 = 1$. It follows that P is not a *Boolean lattice*. We note, however, that the *sublattice*, P', consisting of the relation $|$ defined on the set $\{30, 15, 10, 6, 5, 3, 2, 1\}$ is a *Boolean lattice* and is *isomorphic* to the *Boolean lattice* consisting of $\mathscr{P}(\{a, b, c\})$ with ordering relation $\subset$ (cf. Stone's Theorem, p. 211).

Note. (i) Alternatively, one can define a *Boolean algebra* to be a set L, containing at least two elements O and I, which admits two binary operations $\vee$, $\wedge$ and one unary operation $'$ such that for all elements, $a, b \in L$:

(*a*) a', $a \wedge b$ and $a \vee b$ are elements of L;

(*b*) $a \vee b = b \vee a, \quad a \wedge b = b \wedge a$;

(*c*) $a \vee (b \wedge c) = (a \vee b) \wedge (a \vee c)$,
$a \wedge (b \vee c) = (a \wedge b) \vee (a \wedge c)$;

(*d*) $a \vee O = a, \quad a \wedge I = a$;

(*e*) $a \vee a' = I, \quad a \wedge a' = O$.

(ii) When the axioms for a Boolean algebra are presented in the above manner the duality between the operations $\wedge$ and $\vee$ is manifest. Indeed, interchanging $\wedge$ and $\vee$, and O and I simultaneously throughout leaves the axiom system unchanged, i.e. the axiom system is **self dual**. It follows that in any relation (theorem) derived from this axiom system the interchange of $\wedge$ and $\vee$ and of O and I simultaneously gives a relation which is also derivable from the axioms.

S is a **Boolean subalgebra** of the Boolean algebra L if S is closed under the three operations $\wedge$, $\vee$ and $'$ (i.e. if whenever $a, b \in S$, then $a \wedge b$, $a \vee b$ and a' are elements of S).

The intersection of all subalgebras containing a given set of elements $a_1, \ldots, a_n$ is said to be the subalgebra **generated** by these elements.

If a Boolean algebra, L, has n generators $a_1, \ldots, a_n$ then it can be shown that every element, a, of L can be expressed as the *join* of one or more elements of the form

$$a_1^{\epsilon_1} \wedge a_2^{\epsilon_2} \wedge \ldots \wedge a_n^{\epsilon_n},$$

where $a_i^{\epsilon_i}$ is either a_i or its complement a_i'.

We then say that the resulting expression is a **disjunctive normal form** for a (**disjunctive canonical form**).

Dually (see note (ii) above), we can express a as the *meet* of one or more terms of the form $a_1^{\epsilon_1} \vee a_2^{\epsilon_2} \vee \ldots \vee a_n^{\epsilon_n}$. The resulting expression is a **conjunctive normal form** for a.

Example. The Boolean algebra P' defined in the example above is *generated* by the elements 2, 3, and 5 (having *complements* 15, 10 and 6 respectively). A *disjunctive normal form* for the element 15 is

$$(15 \wedge 3 \wedge 6) \vee (15 \wedge 10 \wedge 5)\ (=(2' \wedge 3 \wedge 5') \vee (2' \wedge 3' \wedge 5)),$$

a *conjunctive normal form* for 15 is

$$(15 \vee 3 \vee 5) \quad (=(2' \vee 3 \vee 5)).$$

Algebras of sets

A **ring** (or **Boolean ring**) **of sets** is a non-empty family, $\mathscr{R}$, of subsets of a given set X (the *universal* set) such that

$$A, B \in \mathscr{R} \Rightarrow A \cap B \in \mathscr{R} \quad \text{and} \quad A \triangle B \in \mathscr{R},$$

i.e. $\mathscr{R}$ is closed under the formation of intersections and symmetric differences (p. 10).

Note. (i) Alternative, equivalent definitions ask that $\mathscr{R}$ should be closed under the formation of unions and symmetric differences, or unions and differences (p. 10).

(ii) If $\mathscr{R}$ is a ring of sets then it satisfies the axioms for an algebraic ring (p. 30) with respect to the operations of symmetric difference (addition) and intersection (multiplication). Algebraic rings in which every element is multiplicatively *idempotent* (i.e. in which $a.a = a$ for all elements a) are also called Boolean rings.

An **algebra** (**Boolean algebra** or **field**) **of sets** is any family of subsets of a set X which is a ring and contains X.

Note. A ring is an algebra if and only if it is closed under the operation of taking the complement.

A ring $\mathscr{R}$ is called a **σ-ring** if it is closed under the formation of countable unions, i.e., if

$$A_i \in \mathscr{R} \quad (\text{all } i \in I, \text{ a countable set}) \Rightarrow \bigcup_{i \in I} A_i \in \mathscr{R}.$$

A σ-ring will also be closed under countable intersections.

An algebra of sets which is closed under countable unions is known as a **σ-algebra, σ-field** or **Borel field.**

A non-empty family of subsets, $\mathcal{M}$, is said to be a **monotone class** whenever, given any **monotone sequence** of sets (E_n) in $\mathcal{M}$ (i.e. a sequence for which either $E_n \supset E_{n+1}$ for all $n \in \mathbb{N}$, or $E_n \subset E_{n+1}$ for all n) we have $\lim E_n \in \mathcal{M}$. (Here, $\lim E_n = \bigcup_{n=1}^{\infty} E_n$ if (E_n) is increasing and $\lim E_n = \bigcap_{n=1}^{\infty} E_n$ if (E_n) is decreasing.)

Every σ-ring is a monotone class and every monotone ring is a σ-ring.

Example. Let X be any uncountable set (p. 23) and $\mathcal{R}_1$ the family of countable subsets of X. Then $\mathcal{R}_1$ is a *Boolean ring of sets* but not a *Boolean algebra of sets*. $\mathcal{R}_2$, the family of all subsets of X which are countable or have a countable complement, forms a *Boolean algebra of sets*. Since the union of a countable number of countable sets is countable, it follows that $\mathcal{R}_1$ is a σ*-ring* and $\mathcal{R}_2$ a σ*-algebra*. $\mathcal{R}_1$ and $\mathcal{R}_2$ are, *a fortiori*, *monotone classes*.

See also: De Morgan's Laws (p. 202), Stone's Theorem (p. 211), Zorn's Lemma (p. 214).

16 Ordinal numbers

We say that a simply-ordered set $(X, \leqslant)$ (p. 75) is **well-ordered** (cf. p. 21) if every non-empty subset of X has a 'first' element, i.e. if, given $S(\neq \varnothing) \subset X$, there exists $s \in S$ such that $s \leqslant x$ for all $x \in S$.

A **well-ordering** of a set X is the result of defining a binary relation $\leqslant$ on X to itself in such a way that X becomes well-ordered with respect to $\leqslant$.

Zermelo's *Well-Ordering Theorem* states that if X is any set whatsoever, then there exists a well-ordering of X. (The Well-Ordering Theorem is equivalent to the Axiom of Choice, p. 201.)

Two well-ordered sets which are isomorphic are said to have the same **order type.**

An order type of particular interest, which we shall denote by ω, is that which contains sets satisfying the axioms

(i) X is simply ordered (p. 75) with respect to $\leqslant$,

(ii) $\forall a \in X$, the set $\{x \in X | x < a\}$ is finite,

(iii) X has no 'last' element, i.e., given any element $a \in X$, $\exists b \in X$ such that $a < b$.

It can be shown that these three axioms form a categorical system (p. 6) and so determine a unique order type.

The order types of well-ordered sets are called **ordinal numbers.**

The ordinal number of a well-ordered set $(W, \leqslant)$ is the order type to which W belongs – it is often denoted by $\overline{W}$.

Note. (i) ω as defined above is an ordinal number, and $\overline{(\mathbb{N}, \leqslant)} = \omega$.

(ii) A finite set, when simply ordered, belongs to a uniquely defined order type determined solely by the cardinality (p. 21) of the set. We may, therefore, use the cardinal number of a finite set to denote its ordinal number. An infinite set, however, may be so ordered as to belong to more than one well-ordering type. For that reason we do not speak of a set having ordinal number $\aleph_0$, say. It is for this reason that the order type ω was introduced above as the order type of the natural numbers in their natural order.

The **sum** $\alpha+\beta$ of the two ordinal numbers α, β represented by (X, R) and (Y, S) respectively $(X \cap Y = \varnothing)$, is defined to be the order type of the well-ordered set $(X \cup Y, T)$ where T is defined so that every element of Y 'follows' every element of X, i.e. by

$$x_1 T x_2 \Leftrightarrow x_1 R x_2 \quad \text{for all} \quad x_1, x_2 \in X,$$
$$y_1 T y_2 \Leftrightarrow y_1 S y_2 \quad \text{for all} \quad y_1, y_2 \in Y,$$
$$xTy \quad \text{for all} \quad x \in X, y \in Y.$$

Addition is associative but not commutative.

The **product** $\alpha\beta$ of the two ordinal numbers α, β represented by (X, R) and (Y, S) respectively, is defined to be the order type of the well-ordered set $(X \times Y, T)$ where T is the *lexicographical* ordering defined by

$$(x_1, y_1) T (x_2, y_2) \quad \text{if and only if:}$$

either $\quad x_1 R x_2$ and $x_1 \neq x_2$

or $\quad x_1 = x_2$ and $y_1 S y_2$.

Multiplication is associative but not commutative.

Every collection of ordinal numbers can be well-ordered by setting $\alpha \leqslant \beta$ if and only if there is a subset (*section*) of a set Y, of order type β, which has order type α.

We thus obtain the **ordinal series**

$$0, 1, 2, \ldots; \quad \omega, \omega+1, \ldots, \omega+n, \ldots; \quad 2\omega, 2\omega+1, \ldots; \quad n\omega, \ldots;$$
$$\omega^2, \omega^2+1, \ldots; \quad \omega^2+\omega, \omega^2+\omega+1, \ldots; \ldots; \quad \omega^n, \omega^n+1, \ldots$$

in which every ordinal number is the order type of the well-ordered set of ordinal numbers which precede it.

The ordinals appearing in the series designate order types of *countable* sets (p. 23). There must, therefore, be a first ordinal which is not in this series and which is the order type of an uncountable set. We denote this ordinal by ω_1.

To distinguish between the various types of ordinals we allocate them to different classes, namely,

(*a*) **first class** – those ordinals corresponding to well-ordered sets with finite cardinality,

(*b*) **second class** – those ordinals corresponding to well-ordered sets with cardinal $\aleph_0$, e.g. ω,

(*c*) **third class** – those ordinals corresponding to well-ordered sets having cardinality $\aleph_1$, where $\aleph_1$ is the cardinal of a representative of

order type ω_1 – the 'smallest' ordinal which does not belong to the second class,
and so on.

Examples. $(\mathbb{Z}, \leqslant)$ is *not* well-ordered since it does not possess a 'first' element. $\mathbb{N}$ can be well-ordered in many ways and so give rise to different order types. However, since $\mathbb{N}$ is countable, all the resulting order types will belong to the *second class*. For example,

(i) $(\{0, 1, 2, 3, \ldots\}, \leqslant)$, is of order type ω,
(ii) $(\{1, 2, \ldots, 0\}, \leqslant)$, is of order type $\omega+1$,
(iii) $(\{2, 3, 4, \ldots, 0, 1\}, \leqslant)$ is of order type $\omega+2$,
(iv) $(\{0, 2, 4, \ldots, 1, 3, 5, \ldots\}, \leqslant)$ is of order type $\omega+\omega$,
(v) $(\{0, 4, 8, \ldots, 1, 5, 9, \ldots, 2, 6, 10, \ldots, 3, 7, 11, \ldots\}, \leqslant)$ is of order type $\omega+\omega+\omega+\omega\ (= 4\omega)$,
(vi) $(\{1, 2, 3, 5, \ldots, 4, 6, 9, \ldots, 8, 12, 18, \ldots, 16, 24, 36, \ldots\}, \leqslant)$ in which the order is by the number of prime factors (equal or not) has order type $\omega+\omega+\omega+\ldots(= \omega.\omega = \omega^2)$.

We note that $(\{\pi, 0, 1, 2, \ldots\}, \leqslant)$ and $(\{0, 1, 2, \ldots, \pi\}, \leqslant)$ have different order types. Hence,

$$1+\omega = \omega \neq \omega+1.$$

Similarly, 2ω is the order type of

$$(\{(a, 1), (a, 2), (a, 3), \ldots, (b, 1), (b, 2), (b, 3), \ldots\}, \leqslant)$$

(cf. (iv) above), whereas $\omega 2$ is the order type of

$$(\{(1, a), (1, b), (2, a), (2, b), \ldots\}, \leqslant)$$

(cf. (i) above). Hence,

$$2\omega = \omega+\omega \neq \omega 2 = \omega.$$

17 Eigenvectors and eigenvalues

Let V be a vector space over a field F and $t: V \to V$ be a linear transformation of V into itself, i.e. an endomorphism (p. 34) of V. An **eigenvector (characteristic vector)** of t is a non-zero vector $v \in V$ such that $t(v)$ is a scalar multiple of v. $\lambda \in F$ is an **eigenvalue (characteristic value, latent root)** of t if there exists a non-zero vector $v \in V$ such that

$$t(v) = \lambda v.$$

If λ is an eigenvalue of t, then the set of all $v \in V$ such that

$$t(v) = \lambda v,$$

is a non-trivial subspace of V known as the **eigenspace** of λ.

The set of all eigenvalues of a linear transformation t of a finite-dimensional vector space V is known as the **spectrum** of t.

If V is a finite-dimensional vector space of dimension n, then t can be represented, relative to a basis $b_1, b_2, \ldots, b_n$, by an $n \times n$ matrix A (p. 41).

We define the **eigenvectors** and **eigenvalues** of the matrix A to be those of the corresponding transformation t. Hence,

$$x = (x_1, x_2, \ldots, x_n)^T$$

is an eigenvector of $A = (a_{ij})$ with eigenvalue λ if and only if

$$\lambda x = Ax,$$

i.e.

$$\lambda x_i = \sum_j a_{ij} x_j.$$

The matrix equation $Ax = \lambda x$ can be rewritten in the form

$$(A - \lambda I_n)x = 0.$$

If the rank of $A - \lambda I_n$ is n, then this system of equations is a Cramer system (p. 48) having the unique solution $x = 0$. It is a necessary condition, therefore, for the existence of eigenvalues that

$$|A - \lambda I_n| = 0.$$

The left-hand side of this last equation is a polynomial of degree n in λ with coefficients in F. It is known as the **characteristic poly-**

nomial of the matrix A. The eigenvalues of A will be the zeros of the characteristic polynomial of A.

The characteristic polynomial will have the form

$$(-1)^n\lambda^n+(-1)^{n-1}f_{n-1}\lambda^{n-1}+\ldots-f_1\lambda+f_0,$$

where $\quad f_0 = |A| \quad$ and $\quad f_{n-1} = a_{11}+a_{22}+\ldots+a_{nn}.$

The expression appearing on the right-hand side of this last relation, i.e. the sum of those elements appearing in the *leading* diagonal of A, is called the **trace (spur)** of A and is denoted by $\operatorname{tr}(A)$, $\operatorname{Tr}(A)$ or $\operatorname{sp}(A)$.

Example. Let $t: \mathbb{R}^3 \to \mathbb{R}^3$ be described relative to the standard basis e_1, e_2, e_3 of $\mathbb{R}^3$ (see example, p. 45) by

$$A = \begin{pmatrix} 0 & 1 & 0 \\ 0 & 0 & 1 \\ 6 & -11 & 6 \end{pmatrix}.$$

Then the *eigenvectors* of t are vectors (x_1, x_2, x_3) such that

$$\begin{pmatrix} 0 & 1 & 0 \\ 0 & 0 & 1 \\ 6 & -11 & 6 \end{pmatrix}\begin{pmatrix} x_1 \\ x_2 \\ x_3 \end{pmatrix} = \lambda\begin{pmatrix} x_1 \\ x_2 \\ x_3 \end{pmatrix},$$

or

$$\begin{pmatrix} -\lambda & 1 & 0 \\ 0 & -\lambda & 1 \\ 6 & -11 & 6-\lambda \end{pmatrix}\begin{pmatrix} x_1 \\ x_2 \\ x_3 \end{pmatrix} = 0 \quad \text{for some } \lambda \in \mathbb{R}.$$

This last equation has a solution other than $x = 0$ only when the determinant of the 3×3 matrix appearing in it (i.e. $-(\lambda^3-6\lambda^2+11\lambda-6)$) is zero. The polynomial $-(\lambda^3-6\lambda^2+11\lambda-6)$ is the *characteristic polynomial* of A and we note that $A^3-6A^2+11A-6I_3 = 0$ (see Cayley–Hamilton Theorem, p. 200). The zeros of this polynomial, namely 1, 2 and 3, are the *eigenvalues* of t (and A). Thus the *spectrum* of t is $\{1, 2, 3\}$. The *eigenspace* of 1 is the set of all *eigenvectors* of the form $\mu(1, 1, 1)$ where $\mu \in \mathbb{R}$, the *eigenspace* of 2 is the set of *eigenvectors* of the form $\mu(1, 2, 4)$, and that of 3 is the set of *eigenvectors* of the form $\mu(1, 3, 9)$. (We note that t (A) has all its eigenvalues in $\mathbb{R}$. This will not be the case in general since $\mathbb{R}$ is not algebraically closed (p. 72) and so will not necessarily contain the zeros of the characteristic polynomial.)

The *trace* of A, $\operatorname{tr}(A)$, is $0+0+6 = 6$.

A linear transformation $t: V \to V$ is said to be **triangulable** if there exists a basis of V with respect to which the matrix of t is triangular (p. 49).

Similarly, a matrix $A \in M_n(F)$ is **triangulable over** F if there exists a non-singular matrix $P \in M_n(F)$, i.e. $P \in GL(n, F)$, such that PAP^{-1} is triangular.

Triangulability is equivalent to the endomorphism (or matrix) having all its eigenvalues in F.

A linear transformation $t: V \to V$ is said to be **diagonalisable** if there exists a basis of V with respect to which the matrix of t is a diagonal matrix.

Note. The endomorphism t (the matrix A) is diagonalisable if and only if the eigenvectors of t (of A) span (p. 40) the space V (the space F^n).

Example. t and A as defined in the above example have distinct (simple) *eigenvalues* in $\mathbb{R}$ and are therefore *diagonalisable* (and, hence, *triangulable*) in $\mathbb{R}$. Relative to the basis (1, 1, 1), (1, 2, 4), (1, 3, 9) t is described by the matrix

$$B = \begin{pmatrix} 1 & 0 & 0 \\ 0 & 2 & 0 \\ 0 & 0 & 3 \end{pmatrix}.$$

The matrix representing the linear transformation which maps the above basis onto e_1, e_2, e_3 (relative to e_1, e_2, e_3) is

$$C = \begin{pmatrix} 3 & -\frac{5}{2} & \frac{1}{2} \\ -3 & 4 & -1 \\ 1 & -\frac{3}{2} & \frac{1}{2} \end{pmatrix}.$$

We note that

$$C^{-1} = \begin{pmatrix} 1 & 1 & 1 \\ 1 & 2 & 3 \\ 1 & 4 & 9 \end{pmatrix},$$

$C^{-1}BC = A$ and $B = CAC^{-1}$, i.e. A is similar (p. 49) to the diagonal matrix B. Note that the set of eigenvalues of B is equal to that of A (this must, of course, always be the case for similar matrices if the eigenvalues of t are not to depend upon the reference basis chosen for A).

A linear transformation $t: V \to V$ is said to be **nilpotent** when $t^k = 0$ for some positive integer k, i.e. $t^k(v) = 0$ for all $v \in V$. Similarly, a matrix is nilpotent if it represents a nilpotent linear transformation (that is, if there exists a positive integer k for which $A^k = 0$).

An $n \times n$ matrix is said to be **reduced** (or an **elementary Jordan matrix**) if it has the form

$$\begin{pmatrix} \lambda & 1 & 0 & 0 & \dots & 0 \\ 0 & \lambda & 1 & 0 & \dots & 0 \\ 0 & 0 & \lambda & 1 & \dots & 0 \\ \vdots & & & & & \vdots \\ 0 & 0 & 0 & 0 & \dots & \lambda \end{pmatrix},$$

i.e.
$$a_{ii} = \lambda \quad (i = 1, \ldots, n),$$
$$a_{i(i+1)} = 1 \quad (i = 1, \ldots, n-1),$$
$$a_{ij} = 0 \quad (j \neq i, i+1),$$
or if its transpose has that form. (See Jordan Canonical Form Theorem (p. 206).)

Example. The matrix
$$A = \begin{pmatrix} 1 & 1 & 1 \\ -1 & -1 & -1 \\ 1 & 1 & 0 \end{pmatrix}$$
is *nilpotent* since $A^3 = 0$. It has characteristic polynomial $-\lambda^3$ and so has the single eigenvalue 0 with multiplicity 3. It is, therefore, *triangulable* but not *diagonalisable*, and its *Jordan canonical form* is
$$\begin{pmatrix} 0 & 1 & 0 \\ 0 & 0 & 1 \\ 0 & 0 & 0 \end{pmatrix} = \begin{pmatrix} 1 & 0 & -1 \\ 0 & 0 & 1 \\ 1 & 1 & 0 \end{pmatrix} \begin{pmatrix} 1 & 1 & 1 \\ -1 & -1 & -1 \\ 1 & 1 & 0 \end{pmatrix} \begin{pmatrix} 1 & 0 & -1 \\ 0 & 0 & 1 \\ 1 & 1 & 0 \end{pmatrix}^{-1}.$$

See also: Cayley–Hamilton Theorem (p. 200); Jordan canonical form (p. 206).

18 Quadratic forms and inner products

Earlier we defined a *linear form* (p. 44) on a vector space V to be a linear map of V into its ground field F, and a *bilinear form* on V to be a bilinear mapping of $V \times V$ to F (p. 54).

This latter definition can be extended to include the case where f, the bilinear form, is a bilinear mapping from $U \times V$ to F where U and V are *different* vector spaces over F.

A bilinear form $f: V \times V \to F$ is said to be
(*a*) **symmetric** when $f(x, y) = f(y, x)$, for all $x, y \in V$,
(*b*) **skew-symmetric** when $f(x, y) + f(y, x) = 0$ for all $x, y \in V$.

Note. The form f can be uniquely described, relative to a preassigned basis $b_1, \ldots, b_n$ of V, by the matrix $A = (a_{ij})$ given by

$$a_{ij} = f(b_i, b_j).$$

It follows that a form will be symmetric if and only if any matrix A representing it satisfies $A = A^T$ (i.e., is *symmetric* by the definition on p. 43). We correspondingly define a matrix to be *skew-symmetric* if $A^T = -A$.

Let V be a vector space of dimension n over some field F having characteristic (p. 31) other than 2. We define a **quadratic form,** q, on V to be a mapping $V \to F$ given by an expression of the form

$$q(x) = \sum_{ij} k_{ij} x_i x_j,$$

where k_{ij} are fixed elements of F and $x_1, \ldots, x_n$ are the coordinates of x with respect to some preassigned basis. (q is thus associated with a homogeneous polynomial of degree 2 (p. 56).)

Note. (*a*) $q(-x) = q(x)$ for all $x \in V$,
(*b*) $f: (x, y) \mapsto \frac{1}{2}(q(x+y) - q(x) - q(y))$ defines a symmetric bilinear form

$$f: V \times V \to F.$$

An alternative definition of a *quadratic form* is a mapping $q: V \to F$ which satisfies (*a*) and (*b*).

Every symmetric bilinear form will arise from a unique quadratic form in the way that we obtained f from q.

The expression $$q(x) = \sum_{ij} k_{ij} x_i x_j$$

can be rewritten in the form

$$q(x) = x^T A x,$$

where x denotes the column matrix with coordinates $x_1, \ldots, x_n$ and

$$A = (a_{ij}) \in M_n(F) \, (a_{ii} = k_{ii} \, (i = 1, \ldots, n), \quad a_{ij} = \tfrac{1}{2}(k_{ij} + k_{ji}) \, (i \neq j)).$$

The symmetric matrix A is known as the **matrix of the quadratic form** with respect to the preassigned basis of V.

It can be shown that if q is a quadratic form on a vector space of dimension n over $\mathbb{R}$, then by a suitable choice of basis, q can be expressed in the form

$$q(x) = \sum_{i=1}^{n} k_i x_i^2,$$

where

$$k_i = +1, \quad 1 \leqslant i \leqslant s,$$
$$k_i = -1, \quad s+1 \leqslant i \leqslant r,$$
$$k_i = 0, \quad r+1 \leqslant i \leqslant n.$$

r is the **rank** of q, s is known as the **signature** of q.

An alternative definition is that the **signature** of q is $2s-r$.

This form is known as the **canonical form** of the real quadratic form q.

If $r = n$, then q is said to be **non-degenerate.**

A real quadratic form is said to be **positive definite** if, whenever x is non-zero, $q(x) > 0$. It is **positive semidefinite** if the condition is relaxed to

$$x \neq 0 \Rightarrow q(x) \geqslant 0 \quad \text{all} \quad x \in V,$$

and is **negative definite** if

$$x \neq 0 \Rightarrow q(x) < 0 \quad \text{all} \quad x \in V.$$

A real symmetric matrix is said to be positive definite, etc., according to whether or not the associated quadratic form is.

Two real symmetric matrices A and B are said to be **congruent** if there is a non-singular real matrix P such that

$$B = PAP^T.$$

Note. If A and B are congruent, then so are the bilinear forms f and g determined by them, in the sense that there is a linear transformation t mapping V onto V such that

$$f(x, y) = g(t(x), t(y)) \quad \text{for all} \quad x, y \in V.$$

Example. The function $f: \mathbb{R}^3 \times \mathbb{R}^3 \to \mathbb{R}$ which maps (a, b) onto

$$(a_1 a_2 a_3)\begin{pmatrix} 5 & 2 & 3 \\ 2 & 6 & 4 \\ 3 & 4 & 1 \end{pmatrix}\begin{pmatrix} b_1 \\ b_2 \\ b_3 \end{pmatrix}$$

(where (a_1, a_2, a_3) and (b_1, b_2, b_3) are the components of a and b relative to some preassigned basis, e_1, e_2, e_3, say), is a *symmetric bilinear form.* The function $q: \mathbb{R}^3 \to \mathbb{R}$ which maps x onto

$$(x_1 x_2 x_3)\begin{pmatrix} 1 & 2 & -8 \\ 0 & 5 & -3 \\ 4 & 3 & -4 \end{pmatrix}\begin{pmatrix} x_1 \\ x_2 \\ x_3 \end{pmatrix}$$

is a *quadratic form* which corresponds to the *symmetric bilinear form* with matrix

$$A = \begin{pmatrix} 1 & 1 & -2 \\ 1 & 5 & 0 \\ -2 & 0 & -4 \end{pmatrix}.$$

q may then be defined by $q(x) = x^T Ax$, A being the *matrix of the quadratic form* q.

Alternatively, we may write

$$\begin{aligned} q(x) &= x_1^2 + 5x_2^2 - 4x_3^2 + 2x_1x_2 - 4x_1x_3 \\ &= (x_1 + (x_2 - 2x_3))^2 + 4(x_2 + \tfrac{1}{2}x_3)^2 - 9x_3^2. \end{aligned}$$

Thus, by a suitable change of basis, q can be presented in the form

$$q(x) = x_1^{*2} + x_2^{*2} - x_3^{*2}$$

where (x_1^*, x_2^*, x_3^*) are the components of x relative to the new basis. This is the *canonical form* for q and from it we see that q has *rank* 3 (and so is *non-degenerate*) and *signature* 2. q is neither *positive definite* nor *negative definite.*

The components (x_1^*, x_2^*, x_3^*) of x relative to the new basis are given by

$$\begin{pmatrix} x_1^* \\ x_2^* \\ x_3^* \end{pmatrix} = \begin{pmatrix} 1 & 1 & -2 \\ 0 & 2 & 1 \\ 0 & 0 & 3 \end{pmatrix}\begin{pmatrix} x_1 \\ x_2 \\ x_3 \end{pmatrix} \quad \text{or} \quad \begin{pmatrix} x_1 \\ x_2 \\ x_3 \end{pmatrix} = \begin{pmatrix} 1 & -\frac{1}{2} & \frac{5}{6} \\ 0 & \frac{1}{2} & -\frac{1}{6} \\ 0 & 0 & \frac{1}{3} \end{pmatrix}\begin{pmatrix} x_1^* \\ x_2^* \\ x_3^* \end{pmatrix}.$$

The new basis will be e_1, $-\frac{1}{2}e_1 + \frac{1}{2}e_2$, $\frac{5}{6}e_1 - \frac{1}{6}e_2 + \frac{1}{3}e_1$.

Substituting in $q(x) = x^T Ax$, we obtain

$$\begin{aligned} q(x) &= (x_1^* x_2^* x_3^*)\begin{pmatrix} 1 & 0 & 0 \\ -\frac{1}{2} & \frac{1}{2} & 0 \\ \frac{5}{6} & -\frac{1}{6} & \frac{1}{3} \end{pmatrix}\begin{pmatrix} 1 & 1 & -2 \\ 1 & 5 & 0 \\ -2 & 0 & -4 \end{pmatrix}\begin{pmatrix} 1 & -\frac{1}{2} & \frac{5}{6} \\ 0 & \frac{1}{2} & -\frac{1}{6} \\ 0 & 0 & \frac{1}{3} \end{pmatrix}\begin{pmatrix} x_1^* \\ x_2^* \\ x_3^* \end{pmatrix} \\ &= (x_1^* x_2^* x_3^*)\begin{pmatrix} 1 & 0 & 0 \\ 0 & 1 & 0 \\ 0 & 0 & -1 \end{pmatrix}\begin{pmatrix} x_1^* \\ x_2^* \\ x_3^* \end{pmatrix}. \end{aligned}$$

The matrices $\begin{pmatrix} 1 & 1 & -2 \\ 1 & 5 & 0 \\ -2 & 0 & -4 \end{pmatrix}$ and $\begin{pmatrix} 1 & 0 & 0 \\ 0 & 1 & 0 \\ 0 & 0 & -1 \end{pmatrix}$

are therefore *congruent*.

A vector space V over $\mathbb{R}$ together with a symmetric bilinear form f whose associated quadratic form is positive definite is said to be an **inner-product space.**

The value $f(u, v) \in \mathbb{R}$ is written $\langle u, v\rangle$ (or, at the risk of some confusion, (u, v)) and is known as the **inner product** of u and v. The inner product has the following properties: for all $u, v \in V$ and $\lambda, \mu \in \mathbb{R}$,

(*a*) $\langle u, v\rangle = \langle v, u\rangle$ (symmetry),

(*b*) $\langle \lambda u_1+\mu u_2, v\rangle = \lambda\langle u_1, v\rangle+\mu\langle u_2, v\rangle$ (bilinearity (with (*a*))),

(*c*) $u \neq 0 \Rightarrow \langle u, u\rangle > 0$ (positive definite).

Note. An *inner product* can be defined to be a mapping $f: V\times V \to \mathbb{R}$ satisfying (*a*), (*b*) and (*c*).

Example. $\mathbb{R}^3$ becomes an *inner-product space* when we define

$$\langle x, y\rangle^* = 4x_1y_1+x_1y_2+2x_1y_3+x_2y_1+x_2y_2+2x_3y_1+2x_3y_3.$$

The **standard inner product** on $\mathbb{R}^n$, however, is

$$\langle x, y\rangle = x_1y_1+x_2y_2+\ldots+x_ny_n$$

(the *scalar product* or the *dot product* of the vectors x and y, see p. 169).

The **length (norm)** of the vector v, denoted by $|v|$ or $\|v\|$ is defined to be $\sqrt{\langle v, v\rangle}$.

Two vectors are said to be **orthogonal,** written $u \perp v$, when $\langle u, v\rangle = 0$.

If W is a subspace of V, then the space of all vectors orthogonal to W, i.e. $\{v \in V | \langle v, w\rangle = 0 \text{ for all } w \in W\}$, is a subspace of V known as the **orthogonal complement** of W and denoted by $W^{\perp}$.

Compare these definitions with those of *orthogonal* and *annihilator* given on p. 44, noting that, given an inner product $\langle u, v\rangle$, one can define a linear mapping $f: V \to V^*$ by $f(b): a \mapsto \langle a, b\rangle$, i.e. $f(b) \in V^*$ is a linear form mapping V into $\mathbb{R}$.

The set of vectors $\{v_1, \ldots, v_k\}$ is said to be **orthogonal** if

$$\|v_i\| \neq 0 \; (i = 1, \ldots, k\} \quad \text{and} \quad \langle v_i, v_j\rangle = 0 \; (i \neq j).$$

If $k = n$, the dimension of V, then the set of vectors forms an **orthogonal basis** for V. If, in addition, $\|v_i\| = 1$ $(i = 1, \ldots, k)$, then we say that the set (basis) is **orthonormal.**

A linear transformation $t: U \to U'$, where U and U' are inner-product spaces, is **orthogonal** when

$$\langle t(u), t(v)\rangle = \langle u, v\rangle \quad (\text{all } u, v \in U).$$

An $n \times n$ matrix $A \in M_n(\mathbb{R})$ is **orthogonal** when its columns (considered as vectors of $\mathbb{R}^n$) are orthonormal with respect to the standard inner product on $\mathbb{R}^n$ (see example above), i.e. when $A^T A = AA^T = I_n$.

A transformation $U \to U$ is orthogonal when its matrix relative to an orthonormal basis is.

Example. Relative to the standard inner product on $\mathbb{R}^3$ the vectors

$$e_1 = (1, 0, 0), \quad e_2 = (0, 1, 0) \quad \text{and} \quad e_3 = (0, 0, 1)$$

have *unit length* and are pairwise *orthogonal*. (However, relative to the inner product space defined by $\langle\ \rangle^*$ above, we have

$$\langle e_1, e_1\rangle^* = 4, \quad \text{i.e. } \|e_1\| = 2,$$

and $$\langle e_1, e_2\rangle^* = 1, \quad \text{i.e. } e_1 \text{ and } e_2 \text{ are not } \textit{orthogonal}.)$$

The *orthogonal complement* of the subspace of $\mathbb{R}^3$ generated by the vector $(1, 1, 1)$ is the subspace spanned by the vectors $(1, 0, -1)$ and $(0, 1, -1)$. The vectors e_1, e_2, e_3 form an *orthonormal basis* for $\mathbb{R}^3$, the vectors $(1, 1, 1)$, $(1, 0, -1)$ and $(0, 1, -1)$ form a *basis* for $\mathbb{R}^3$ but *not* an *orthogonal basis*, the vectors $(1, 1, 1)$, $(1, 0, -1)$ and $(1, -2, 1)$ form an *orthogonal basis* for $\mathbb{R}^3$, the vectors $(1/\sqrt{3}, 1/\sqrt{3}, 1/\sqrt{3})$, $(1/\sqrt{2}, 0, -1/\sqrt{2})$ and $(1/\sqrt{6}, -2/\sqrt{6}, 1/\sqrt{6})$ form an *orthonormal* basis for $\mathbb{R}^3$. The linear transformation $t: \mathbb{R}^3 \to \mathbb{R}^3$ defined by

$$e_1 \mapsto (1/\sqrt{3}, 1/\sqrt{3}, 1/\sqrt{3})$$
$$e_2 \mapsto (1/\sqrt{2}, 0, -1/\sqrt{2})$$
$$e_3 \mapsto (1/\sqrt{6}, -2/\sqrt{6}, 1/\sqrt{6})$$

is *orthogonal*. The coordinates (x_1', x_2', x_3') of $t(x)$ are given by

$$\begin{pmatrix} x_1' \\ x_2' \\ x_3' \end{pmatrix} = \begin{pmatrix} 1/\sqrt{3} & 1/\sqrt{2} & 1/\sqrt{6} \\ 1/\sqrt{3} & 0 & -2/\sqrt{6} \\ 1/\sqrt{3} & -1/\sqrt{2} & 1/\sqrt{6} \end{pmatrix} \begin{pmatrix} x_1 \\ x_2 \\ x_3 \end{pmatrix},$$

i.e. $x' = Ax$ where A is an *orthogonal* matrix.
We note that

$$\langle t(x), t(y)\rangle = (x')^T y' = x^T A^T A y = x^T y = \langle x, y\rangle.$$

Given $t: U \to U$, a linear transformation of an inner product space U onto itself, then one can define a linear transformation $t^*: U \to U$ by $t^*: v \mapsto w$ where w is the unique element satisfying

$$\langle t(u), v\rangle = \langle u, w\rangle \quad \text{for all} \quad u \in U.$$

The transformation t^* is known as the **adjoint** of t.

t is said to be **self-adjoint** if $t = t^*$.

t is self-adjoint if and only if its matrix, relative to an orthonormal basis, is symmetric (p. 43).

If $t = -t^*$ then t is said to be **skew-symmetric.**

If t is skew-symmetric, then A, the matrix of t, will satisfy $A^T = -A$, and will be a **skew-symmetric matrix.**

Note. The definitions introduced above can be extended in many ways. The simplest and most important extension is to vector spaces over $\mathbb{C}$, the field of complex numbers. In that particular case we have:

(i) A vector space V over $\mathbb{C}$ together with a mapping $f: V \times V \to \mathbb{C}$, $(u, v) \mapsto \langle u, v \rangle$, satisfying (for all $u, v \in V$, $\lambda, \mu \in \mathbb{C}$)

(*a*) $\langle u, v \rangle = \overline{\langle v, u \rangle}$,

(*b*) $\langle \lambda u_1 + \mu u_2, v \rangle = \lambda \langle u_1, v \rangle + \mu \langle u_2, v \rangle$,

(*c*) $u \neq 0 \Rightarrow \langle u, u \rangle > 0$,

is known as a **unitary** space and the inner product is said to be **Hermitian.** (Note that (*a*) implies that $\langle u, u \rangle \in \mathbb{R}$ and, hence, that (*c*) makes sense.)

(ii) The **standard inner product** of $(x_1, \ldots, x_n)$ and $(y_1, \ldots, y_n)$, vectors of $\mathbb{C}^n$, is

$$\langle x, y \rangle = x_1\bar{y}_1 + x_2\bar{y}_2 + x_3\bar{y}_3 + \ldots + x_n\bar{y}_n.$$

(Note that it will no longer be true that $\langle x, y \rangle = \langle y, x \rangle$ for all x, y.)

(iii) A linear transformation $t: V \to V'$, where V and V' are unitary spaces, is a **unitary transformation** when

$$\langle t(u), t(v) \rangle = \langle u, v \rangle \quad (\text{all } u, v \in V).$$

A matrix $A \in M_n(\mathbb{C})$ is **unitary** when

$$A^T \bar{A} = I_n.$$

(t: $V \to V$ is unitary if and only if the matrix of t relative to an orthonormal basis is unitary.)

(iv) If V is a unitary space, then the adjoint, t^*, of $t: V \to V$ is called the **Hermitian adjoint.** A self-adjoint transformation is called **Hermitian.** If $t = -t^*$ then t is said to be **skew-Hermitian.** $A \in M_n(\mathbb{C})$ is said to be **Hermitian** if $A = \bar{A}^T$ and **skew-Hermitian** if $A = -\bar{A}^T$. (A transformation t is Hermitian if and only if it is represented relative to an orthonormal basis by a Hermitian matrix.)

(v) All eigenvalues of a self-adjoint transformation are real.

(vi) The set of all $n \times n$ unitary matrices under matrix multiplication forms a group, the **unitary group.** (Correspondingly, the set of all $n \times n$ orthogonal matrices (p. 91) forms a group, the **orthogonal group** O_n or $O(n, \mathbb{R})$. Those orthogonal matrices having determinant $+1$ form a subgroup SO_n of O_n, known as the **special orthogonal group.**)

Examples. (i) Let $t: \mathbb{C}^3 \to \mathbb{C}^3$ be defined (relative to a preassigned orthonormal basis) by the matrix

$$A = \begin{pmatrix} 1 & 1+i & 0 \\ 1-i & 0 & 2i \\ 0 & -2i & 4 \end{pmatrix}.$$

Then the components (u_1', u_2', u_3') of the vector $t(u)$ will be given by

$$u' = Au.$$

With respect to the *standard inner product* of $\mathbb{C}^3$ we have

$$\langle u, v\rangle = u^T\bar{v},$$

and hence
$$\langle t(u), v\rangle = u^T A^T \bar{v} = \langle u, w\rangle$$

when we define $\bar{w} = A^T\bar{v}$ or $w = \bar{A}^T v$. Thus $\bar{A}^T$ is the matrix of the *Hermitian adjoint* of t. Since, in this case, $\bar{A}^T = A$, t is *self-adjoint* and A is *Hermitian*. Since $A^T\bar{A} \neq I_3$, A is not unitary. We note that the characteristic polynomial (p. 83) for A is $-\lambda^3+5\lambda^2+2\lambda-12$ and that this cubic has three *real* roots, one satisfying $x < 0$, one satisfying $1 < x < 2$ and one satisfying $x > 2$.

(ii) The transformation $s : \mathbb{C}^3 \to \mathbb{C}^3$ defined by the matrix

$$B = \frac{1}{\sqrt{3}}\begin{pmatrix} 1 & \omega & \omega^2 \\ \omega & 1 & \omega^2 \\ 1 & 1 & 1 \end{pmatrix},$$

where ω denotes a complex cube root of unity (p. 166), is a *unitary transformation*, for

$$\langle s(u), s(v)\rangle = u^T B^T \overline{Bv} = u^T(B^T\bar{B})\bar{v} = u^T\bar{v} = \langle u, v,\rangle$$

and $B^T\bar{B} = I_3$, i.e. B is a *unitary matrix.*

See also: Gram–Schmidt orthogonalization process (p. 203), Principal Axis Theorem (p. 209), Schwarz inequality (p. 210), Sylvester's Law (p. 211).

19 Categories and functors

Categorical algebra represents a fundamental approach to the study of such abstract structures as groups, rings, fields, topological spaces, etc. The emphasis in this approach lies not so much on individual groups or rings, say, but on the connections between similar structures and the functions which relate them. The notion of a category was first introduced in a topological setting by Saunders MacLane (b. 1909) and Samuel Eilenberg (b. 1913) in 1945. Since then the concept has been extended to help unify other aspects of mathematics.

A **category** $\mathscr{C}$ is a class of objects, $A, B, C, \ldots$, together with two functions:

(*a*) a function which assigns to each pair of objects $A, B \in \mathscr{C}$ a set $\mathscr{C}(A, B)$ (or $\hom(A, B)$) called the *set of morphisms* with domain A and codomain B;

(*b*) a function assigning to each triple of objects $A, B, C \in \mathscr{C}$ a law of composition

$$\mathscr{C}(A, B) \times \mathscr{C}(B, C) \to \mathscr{C}(A, C),$$

(if $f \in \mathscr{C}(A, B)$ and $g \in \mathscr{C}(B, C)$, then the composite of f and g written gf, is a morphism with domain A and codomain C).

The functions and morphisms so defined must, moreover, satisfy the axioms:

I $\mathscr{C}(A_1, B_1)$ and $\mathscr{C}(A_2, B_2)$ are disjoint unless $A_1 = A_2$ and $B_1 = B_2$;

II (Associativity) Given $f \in \mathscr{C}(A, B)$, $g \in \mathscr{C}(B, C)$ and $h \in \mathscr{C}(C, D)$, then $(hg)f = h(gf)$;

III (Identity) To each object A in $\mathscr{C}$ there is a morphism $1_A : A \to A$ such that, for all $f \in \mathscr{C}(A, B)$, $g \in \mathscr{C}(C, A)$, we have

$$f1_A = f, \quad 1_A g = g.$$

Note. (i) The word 'class' was used in the first line of the definition rather than 'set' since we wish to consider collections which are too 'large' to be sets, e.g., the collection of all sets. A class A will be called a *set* only when it is itself an element of a class B. The axiom of class formation allows us to form the class of all sets having a certain property (cf. p. 8 where we only considered elements of a *particular set* which possessed a given property and *not* elements in general).

(ii) Some authors take the opportunity offered by this definition to rid themselves of the burden of writing gf to mean the morphism f followed by the morphism g. They denote the composite of f and g as defined above by fg. (See note on p. 14.)

(iii) The morphism 1_A is determined uniquely and is known as the **identity morphism** for A.

In the particular case when, to each of the objects $A, B, C, \ldots$, there corresponds a set $U(A)$, $U(B)$, ... known as the 'underlying set' of $A, B, \ldots$, and the elements of $\mathscr{C}(A, B)$, etc. are functions from the set $U(A)$ to the set $U(B)$, $\mathscr{C}$ is called a **concrete category.**

If the class of objects $A, B, C, \ldots$ forms a set, then $\mathscr{C}$ is called a **small category.**

Examples. Typical *categories* are:

(i) The *category*, $\mathscr{S}$, of sets and functions. Here $\mathscr{C}$ consists of the class of all sets and the set $\mathscr{C}(A, B)$ is the set of functions with domain A and codomain B. The *identity morphism* for A, 1_A, is the identity function $id: A \to A$.

(ii) The *category*, $\mathscr{G}$, of groups and homomorphisms. Here $\mathscr{C}$ consists of the class of all groups and the set $\mathscr{C}(A, B)$ is the set of homomorphisms from A to B. 1_A is the identity automorphism $A \to A$.

(i) and (ii) are examples of *concrete categories*. In (i) the underlying set of A is A itself, in (ii) the underlying set of the group $(G, \circ)$ is the set G.

An example of a *small category* is

(iii) Let $\mathscr{C}$ consist of the group G alone and define $\mathscr{C}(G, G)$ to be the set of right translations $R_g: G \to G$ defined by

$$R_g: x \mapsto xg.$$

In this way we obtain a morphism R_g corresponding to every $g \in G$. The *identity morphism* is the translation R_e where e is the identity element of G.

A morphism $f \in \mathscr{C}(A, B)$ is said to be **invertible** (a **unit** or an **isomorphism**) in $\mathscr{C}$ if there exists $g \in \mathscr{C}(B, A)$ such that

$$gf = 1_A, \quad fg = 1_B.$$

If f is invertible, we write $g = f^{-1}$ and say that A and B are **equivalent** in $\mathscr{C}$. A category in which every morphism is invertible is known as a **groupoid** (but see note on p. 25).

A morphism $m \in \mathscr{C}(B, C)$ is said to be **monic** in $\mathscr{C}$ if, for all $f, f' \in \mathscr{C}(A, B)$ *and* all A in $\mathscr{C}$, $mf = mf'$ implies $f = f'$.

A morphism $e \in \mathscr{C}(A, B)$ is **epic** in $\mathscr{C}$ if $ge = g'e$ for two morphisms $g, g' \in \mathscr{C}(B, C)$ always implies $g = g'$.

Monic and epic morphisms are, therefore, generalisations of mono- and epimorphisms (p. 34).

An object I in $\mathscr{C}$ is said to be an **initial object** in $\mathscr{C}$ if for all X in $\mathscr{C}$ there is exactly one morphism $I \to X$, i.e. if $\mathscr{C}(I, X)$ is a singleton (p. 9) for all X in $\mathscr{C}$. Similarly an object T in $\mathscr{C}$ is called a **terminal object** if $\mathscr{C}(X, T)$ is a singleton for all X in $\mathscr{C}$. An object which is both initial and terminal is known as a **zero object**.

If $\mathscr{C}$ possesses a zero object P, then, for all A, B in $\mathscr{C}$, $\mathscr{C}(A, B)$ contains the morphism gf where $\{f\} = \mathscr{C}(A, P)$ and $\{g\} = \mathscr{C}(P, B)$. gf is independent of the choice of zero object and is known as the **zero morphism**. It is denoted by $0_{AB} : A \to B$.

Examples. Consider the category, $\mathscr{G}$, defined above. Then $f \in \mathscr{C}(A, B)$ is *invertible* if f is bijective, i.e. if f is an isomorphism from A to B. In this case the groups A and B are (isomorphically) *equivalent*. f is *monic* if it is injective (i.e. a monomorphism), it is *epic* if it is surjective (i.e. an epimorphism). The *zero objects* of $\mathscr{G}$ are the trivial groups ($\{e\}$, $\{1\}$, ...). Since the image of the identity element is uniquely determined by a homomorphism, $\mathscr{C}(\{e\}, A)$ is a singleton and $\{e\}$ is an *initial* object. Similarly, $\mathscr{C}(A, \{e\})$ is a singleton and $\{e\}$ is a *terminal* object.

In the *small category* defined in (iii) above, every morphism is *invertible* since $R_{g^{-1}}R_g = R_{g^{-1}}R_g = 1_G$. This category is therefore a *groupoid*.

Given two categories, $\mathscr{C}$ and $\mathscr{D}$, we define a **functor** $F : \mathscr{C} \to \mathscr{D}$ to be a pair of functions, namely,

(*a*) the *object function* that assigns to each A in $\mathscr{C}$ an object $F(A)$ in $\mathscr{D}$; and

(*b*) the *mapping function* that assigns to each morphism $f : A \to B$ of $\mathscr{C}$ a morphism $F(f) : F(A) \to F(B)$ of $\mathscr{D}$, which satisfy

I $F(1_A) = 1_{F(A)}$ for each A in $\mathscr{C}$,

II $F(gf) = F(g)F(f)$ for each composite gf defined in $\mathscr{C}$.

Note. A functor can be thought of as a morphism of categories. It is possible to define an *isomorphism* of categories and to talk of *invertible functors* and *equivalent categories*.

The definition given above is strictly that of a **covariant functor.** If one asks in (*b*) that the mapping function should assign to each $f : A \to B$ of $\mathscr{C}$ a morphism $F(f) : F(B) \to F(A)$ and, moreover, that II should then be replaced by

$$F(gf) = F(f)F(g),$$

then F is known as a **contravariant functor.**

Examples. (i) Given a *concrete category*, $\mathscr{P}$, one can define an *underlying set functor* U from $\mathscr{P}$ to the category of underlying sets $\mathscr{S}$, by taking the *object function* to be the function which maps $P \in \mathscr{P}$ onto its underlying set in $\mathscr{S}$, and the *mapping function* to be the function which assigns to each morphism

$f: A \to B$ (in $\mathscr{P}$) the set function $f: A \to B$ (in $\mathscr{S}$). The effect of the functor is to throw away any algebraic or topological structure there might be on $\mathscr{P}$, and for this reason U is said to be a *forgetful functor*.

(ii) Let G and H be groups. Then we can define a *functor* from the category determined by G (example p. 95) to that determined by H, by mapping G onto H (*object function*) and R_g onto $R_{\phi(g)}$ (*mapping function*) where ϕ is any homomorphism from G to H.

(iii) We can construct a second category $\mathscr{G}^{\text{op}}$ from $\mathscr{G}$ (example p. 95) by taking $\mathscr{G}^{\text{op}}$ to comprise the class of all groups together with morphisms of the type $f^{\text{op}}: H \to G$, each f^{op} corresponding to a morphism $f: G \to H$ in $\mathscr{G}$. The composite $f^{\text{op}}g^{\text{op}} = (gf)^{\text{op}}$ is defined whenever gf is defined and $\mathscr{G}^{\text{op}}$ satisfies all the conditions for a category. The two functions $G \mapsto G$ (*object function*) and $f \mapsto f^{\text{op}}$ (*mapping function*) define a *contravariant functor* from $\mathscr{G}$ to $\mathscr{G}^{\text{op}}$.

Given two functors $F, G: \mathscr{C} \to \mathscr{D}$ then a **natural transformation** $: F \to G$ is a function which assigns to each object $X \in \mathscr{C}$ a morphism $t_X: F(X) \to G(X)$ in $\mathscr{D}$ in such a way that every morphism $f: X \to Y$ of $\mathscr{C}$ yields a commutative diagram, i.e. that

$$G(f)\, t_X = t_Y\, F(f).$$

If t_X is invertible for each X, then t is called a **natural isomorphism** or **natural equivalence**.

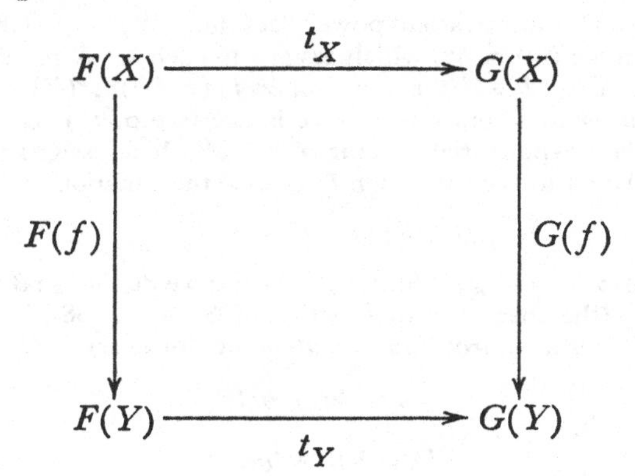

Note. Natural transformations were developed in order to bring precision to the intuitive idea of a 'natural map' (see examples below). They can be thought of as morphisms of functors.

Examples (i). Let Ab be defined on the class of all groups by

$$Ab(G) = G_1 = G/[G, G]$$

(p. 60). (G_1 will be an Abelian group.) Any homomorphism $\phi : G \to H$ maps $[G, G]$ into $[H, H]$ and induces a homomorphism

$$\phi'(= Ab\phi) : G_1 \to H_1.$$

Hence, *Ab* is a *functor* from $\mathscr{G}$ to the *subcategory* of Abelian groups and homomorphisms. If we now define $t_G : G \to G_1$ in the obvious manner, then t is a *natural transformation* from the *identity functor* $I : \mathscr{G} \to \mathscr{G}$ to the *Abelianising functor* $Ab : \mathscr{G} \to \mathscr{G}$. We have:

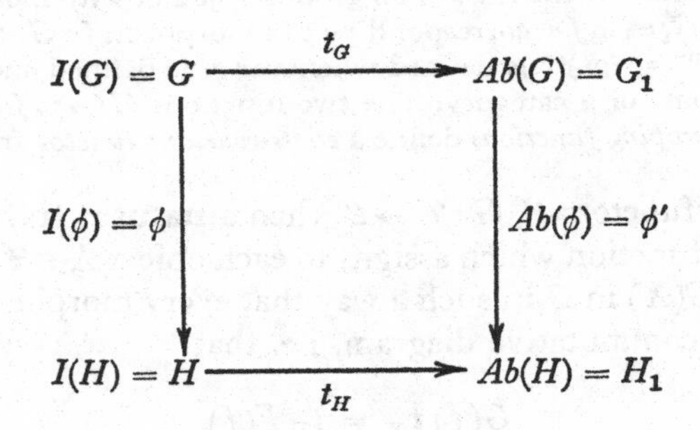

(ii) On p. 13 we drew attention to the 'natural map' between the set $\mathscr{P}(A)$ and the set $\{0, 1\}^A$. This can be described in terms of categorical algebra as follows.

Let $\mathscr{P}$ denote the contravariant power-set functor $\mathscr{S}^{op} \to \mathscr{S}$ (see example (i) p. 95 and example (iii) p. 97) which assigns to each set X the set $\mathscr{P}(X)$, and to each function $f: U \to V$ the function $\mathscr{P}(f)\ (=f^{-1}): \mathscr{P}(V) \to \mathscr{P}(U)$ which maps each subset of V onto its inverse image (see p. 16). Let $h^{\{0,1\}}$ denote the contravariant exponential functor $\mathscr{S}^{op} \to \mathscr{S}$ which assigns to each set X the set $\{0, 1\}^X$ and to each function $f: U \to V$ the function

$$\{0, 1\}^f : \{0, 1\}^V \to \{0, 1\}^U$$

defined by $g(\in \{0, 1\}^V) \mapsto gf(\in \{0, 1\}^U)$. Then if we define $t_X : \mathscr{P}(X) \to \{0, 1\}^X$ by $t_X(S) = \chi_S$ (the characteristic function of S, see p. 188), it can be shown that $t: \mathscr{P} \to h^{\{0,1\}}$ is a *natural transformation*, i.e. for every $f: U \to V$ we have

$$\{0, 1\}^f\, t_V = t_U f^{-1}$$

(for all $X \in \mathscr{P}(V)$,

$$\{0, 1\}^f\, (t_V(X)) = t_U(f^{-1}(X))).$$

20 Metric spaces and continuity

Let E be a non-empty set. We define a **metric** (or **distance**) on E to be a function $d: E \times E \to \mathbb{R}$ satisfying:

I $d(x, y) \geqslant 0$ for all $x, y \in E$;
II $d(x, y) = 0 \Leftrightarrow x = y$;
III $d(x, y) = d(y, x)$ for all $x, y \in E$;
IV for all $x, y, z \in E$,

$$d(x, z) \leqslant d(x, y) + d(y, z) \quad \text{(the \textit{triangle inequality}, cf. p. 69).}$$

A **metric space**, (E, d), is a set E together with a metric d on E.

Note. The notation (E, d) is necessary since two different metrics, d_1 and d_2 say, could be defined upon the same set E and we should then have the two distinct spaces (E, d_1) and (E, d_2). However, in many cases, where no ambiguity arises, one refers to the metric space E. In particular, $\mathbb{R}^n$ (or $\mathbb{R}$ in the case $n = 1$) is used to denote the set $\mathbb{R}^n$ with the Euclidean metric (see example (i) below).

If A is a subset of E, then (A, d) is called a **subspace** of (E, d). We then say that the metric on A is **induced** by that on E. The induced metric might not, however, be the most obvious metric on A (see example (i) below).

Examples. (i) The **Euclidean** (or **Pythagorean**) metric is defined on $\mathbb{R}^n$ by

$$d(x, y) = \sqrt{\{(x_1 - y_1)^2 + (x_2 - y_2)^2 + \ldots + (x_n - y_n)^2\}}.$$

This metric is *induced* on the $(n-1)$ unit sphere $\mathbb{S}^{n-1}$, i.e. the set

$$\{x \in \mathbb{R}^n \mid d(x, 0) = 1\}.$$

The metric which we use, however, when measuring distances on the earth ($\mathbb{S}^2$) is not that induced by the metric on $\mathbb{R}^3$, but the *intrinsic* metric defined to be arc length measured along the great circle through x and y.

(ii) Other metrics of interest include:

(*a*) the **discrete** metric defined by

$$d(x, y) = 1 \quad \text{if} \quad x \neq y;$$

(*b*) the **norm** metric (p. 90) defined on an inner product space by

$$d(x, y) = \|x - y\| \; (= \sqrt{\langle x - y, x - y \rangle});$$

(*c*) the **product** metric defined on the Cartesian product, $A \times B$, of the two metric spaces (A, p) and (B, q) by

$$d((a_1, b_1), (a_2, b_2)) = \sqrt{\{p(a_1, a_2)^2 + q(b_1, b_2)^2\}}.$$

If (E_1, d_1) and (E_2, d_2) are two metric spaces and $f: E_1 \to E_2$ is a bijection satisfying

$$d_2(f(x), f(y)) = d_1(x, y), \quad \text{for all } x, y \in E_1,$$

then we say that f is an **isometry** and that (E_1, d_1) and (E_2, d_2) are **isometric.**

If f is an injection, then we say that f is an **isometric embedding** of E_1 into E_2.

If (E_1, d_1) is a metric space and $f: E_1 \to E_2$ is a bijection, then we can *define* a metric on E_2 by

$$d(x, y) = d_1(f^{-1}(x), f^{-1}(y)).$$

The metric d is then said to have been **transported** from E_1 to E_2 by f.

Open and closed sets

Given a metric space (E, d) and a point $a \in E$, we define the **(open) r-ball** (or **open sphere** – **open disc** if the metric space is a subset of $\mathbb{R}^2$) with **centre a** (and **radius r**) to be $\{x \in E \mid d(x, a) < r\}$ and denote it by $B(a; r)$ or $B_a(r)$.

Similarly, we define the **closed r-ball** (or **closed sphere**) with **centre a** (and **radius r**) and the **sphere** with **centre a** and **radius r** to be

$$\bar{B}(a; r) = \{x \in E \mid d(x, a) \leqslant r\} \text{ (or, alternatively, } B'(a, r)),$$

$$S(a; r) = \{x \in E \mid d(x, a) = r\}.$$

If X is a non-empty subset of E, then we define the **diameter** of X to be

$$\text{lub}\{d(x, y) \mid x, y \in X\} \; (= \sup\{d(x, y) \mid x, y \in X\}).$$

If the diameter is finite, then we say that X **is bounded.**

If X and Y are two non-empty subsets of E, then we define the **distance from X to Y** to be

$$\text{glb}\{d(x, y) \mid x \in X, y \in Y\} \; (= \inf\{d(x, y) \mid x \in X, y \in Y\}).$$

Note. (i) The distance so defined is not a metric on $\mathscr{P}(E)$.
(ii) The **distance of a point x from a non-empty set** Y is defined to be

$$\text{glb}\{d(x, y) \mid y \in Y\}, \quad \text{i.e. the distance of } \{x\} \text{ from } Y.$$

A subset X of a metric space (E, d) is said to be **open** if for each point x in X there is an $r > 0$ such that the open ball $B(x, r) \subset X$.

Note. $\varnothing$ and E itself are always open sets.

A subset Y of a metric space E is said to be **closed** if $E - Y$ is open.

An alternative definition of a closed set can be found on p. 102.

An **open neighbourhood** of a non-empty subset X of E is an open set containing X, a **neighbourhood** of X is any set containing an open neighbourhood of X.

A neighbourhood of a point x is a neighbourhood of the *set* $\{x\}$.

A **fundamental system of neighbourhoods** of X is a set, $\mathcal{N}$, of neighbourhoods of X such that any given neighbourhood of X contains a member of $\mathcal{N}$.

Example. Let $E = \{x_1, x_2\}$ and let d be defined by $d(x_1, x_2) = 1$. Then (E, d) is a *metric space*. (The mapping $f: E \to \mathbb{R}$ defined by $x_1 \mapsto 0$, $x_2 \mapsto 1$ is an *isometric embedding* of E in $\mathbb{R}$ (with the usual Euclidean metric).) The *open ball*, $B(x_1; 1)$, is the set $\{x_1\}$. The *closed ball* $\bar{B}(x_1; 1)$ is the set E. (In $\mathbb{R}$ the *open ball* $B(0; 1)$ is the set $\{x \in \mathbb{R} \mid -1 < x < 1\}$, and the closed ball $\bar{B}(0; 1)$ is the set $\{x \in \mathbb{R} \mid -1 \leqslant x \leqslant 1\}$.) The *diameter* of $\{x_1, x_2\}$ is 1. $\{x_1, x_2\}$ is therefore *bounded* (as is $\{0, 1\}$ in $\mathbb{R}$). The set $\{x_1\}$ is *open* in E, since $B(x_1; 1) \subset \{x_1\}$. (The set $\{0\}$ is *not* open in $\mathbb{R}$, since there is no $r > 0$ such that $B(0; r) \subset \{0\}$.) The set $\{x_1\}$ is *closed* in E since $E - \{x_1\} = \{x_2\}$ is *open*. (The set $\{0\}$ is *closed* in $\mathbb{R}$ since the set $\mathbb{R} - \{0\}$ is *open*.) $\{x_1\}$ is an *open neighbourhood* of x_1 in E. ($\{0\}$ is *not* an *open neighbourhood* of 0 in $\mathbb{R}$, but $B(0, \frac{1}{10})$ is.) $\{\{x_1\}\}$ is a *fundamental system of neighbourhoods* of x_1 in E. ($\{B(0; 1/n) \mid n \in \mathbb{N} - \{0\}\}$ is a *fundamental system of neighbourhoods* of 0 in $\mathbb{R}$.)

x is said to be **interior** to X if X is a neighbourhood of x. The set of all points interior to X is called the **interior** of X and is written $\mathring{X}$ or $\mathscr{I}(X)$.

Alternatively, $\mathring{X}$ is the union of all open sets contained in X.

Given a subset $X \subset E$, a point $c \in E$ is called a **cluster point** of X if every neighbourhood of c has a non-empty intersection with X. If, moreover, every neighbourhood of c meets X in a point other than c, then we say that c is a **limit point** (or **point of accumulation**) of X. If, however, there is a neighbourhood V of c such that $V \cap X = \{c\}$, then c is said to be an **isolated point** of X.

Note. (i) The set of cluster points of X is partitioned into the set of limit points of X and the set of isolated points of X.

(ii) Some authors use *limit point* where we have used *cluster point*.

(iii) A set X which has no limit point is called a **discrete set.**

Given a non-empty subset of E we define the **closure** of X, written $\bar{X}$, to be the set of all cluster points of X and the **derived set** of X, written X', to be the set of all limit points of X.

Note. (i) $\bar{X} = X \cup X'$ and the isolated points of X are those points of X which do not belong to X'

(ii) $\bar{X}$ is also characterised by being the intersection of all closed sets containing X or, alternatively, by being the smallest closed set containing X.

(iii) An alternative definition of a *closed set* (cf. p. 101) is a set which contains all its limit points.

A point $a \in E$ is said to be a **frontier point** (or **boundary point**) of X if it is a cluster point of both X and $E - X$. The set of all frontier points of X is called the **frontier** (or **boundary**) of X and is denoted by $\mathrm{Fr}(X)$ or $\mathscr{F}(X)$.

Alternatively, $a \in \mathrm{Fr}(X)$ if every neighbourhood of a has a non-empty intersection with X and $E - X$.

A point a is said to be **exterior** to X if there exists a neighbourhood of a which contains no point of X, i.e. if a is interior to $E - X$.

The interior of X, the frontier of X and the set of points exterior to X are, therefore, three pairwise disjoint sets whose union is E.

Example. In $\mathbb{R}^2$, the *interior* of $\{(x,y) \mid x^2 + y^2 \leqslant 1\}$ is the set $\{(x,y) \mid x^2 + y^2 < 1\}$, the *interior* of $\{(x,y) \mid x^2 + y^2 = 1\}$ is the empty set. The only *limit point* in $\mathbb{R}^2$ of $X = \{(1, 1), (\frac{1}{2}, 1), (\frac{1}{3}, 1), (\frac{1}{4}, 1), \ldots\}$ is the point $(0, 1)$. The set of *cluster points* of X is $X \cup \{(0, 1)\}$. All points of X are *isolated points* of X. The *closure* of X, $\bar{X}$, is the set $X \cup \{(0, 1)\}$ and the *derived set*, X', is $\{(0, 1)\}$. The *frontier* of X, $\mathrm{Fr}(X)$, is the set $X \cup \{(0, 1)\}$, the interior of X, $\mathring{X}$, is empty, and the set of points *exterior* to X is the set $\mathbb{R}^2 - \bar{X}$. The set $\mathbb{N}$ is *discrete* in $\mathbb{R}$, since it possesses no *limit points*. Moreover, like X, all its points are isolated points.

In a metric space E, a set X is said to be **dense** with respect to a set Y if every point of Y is a cluster point of X, i.e. if $Y \subset \bar{X}$.

Alternatively, if, for all $y \in Y$, any neighbourhood of y contains points of X.

X is said to be **everywhere dense** (or, sometimes, **dense**) if it is dense with respect to E.

A metric space E is said to be **separable** if there exists a countable set (p. 23) which is dense in E.

A **condensation point** of a subset X of a metric space E is a point $x \in E$ such that in every neighbourhood of x there is an uncountable set of points of X.

A closed subset X of a metric space E is said to be **perfect** if each point of X is a limit point of X, i.e. if $X = X'$.

Examples. The set $X = \{q \in \mathbb{Q} | 0 \leqslant q \leqslant 1\}$ is *dense with respect to* the subset $Y = \{x | 0 \leqslant x \leqslant 1\}$ of $\mathbb{R}$. The set $\mathbb{Q}^2 = \{(p, q) | p, q \in \mathbb{Q}\}$ is *everywhere dense* in $\mathbb{R}^2$. Since $\mathbb{Q}^2$ is countable it follows that $\mathbb{R}^2$ is *separable*. The Cantor set, C, (see p. 199) is *perfect* in $\mathbb{R}$, it is also **nowhere dense** in $\mathbb{R}$, that is $\mathbb{R} - \overline{C}$ is dense. (π, e) is a *condensation point* in $\mathbb{R}^2$ of the set $Z = \mathbb{R}^2 - \mathbb{Q}^2$.

Continuous functions

Let (E_1, d_1) and (E_2, d_2) be two metric spaces. A function $f: E_1 \rightarrow E_2$ is said to be **continuous at a point** $a \in E_1$ if, for every neighbourhood V of $f(a)$ in E_2, there is a neighbourhood U of a in E_1 such that $f(U) \subset V$.

If f is continuous at every point of E_1, then we say that f is **continuous** in (or on) E_1.

Note. (i) The definition given above is equivalent to 'For every open set V (in E_2) containing $f(a)$ there is an open set U (in E_1) containing a such that $f(U) \subset V$'.

(ii) An equivalent and more traditional definition is that f is **continuous at a point** $a \in E_1$ if, $\forall \epsilon > 0$, $\exists \delta > 0$ such that $d_1(a, x) < \delta$ implies $d_2(f(a), f(x)) < \epsilon$. (See p. 117 for an interpretation of this definition for a function $f: \mathbb{R} \rightarrow \mathbb{R}$.)

(iii) Topologists often restrict the use of the word 'mapping' to describe functions that are continuous. In this book, however, mapping is used as a synonym for function.

A **uniformly continuous function** is a function $f: E_1 \rightarrow E_2$ for which: $\forall \epsilon > 0$, $\exists \delta > 0$ such that $d_2(f(x), f(y)) < \epsilon$ for all $x, y \in E_1$ satisfying $d_1(x, y) < \delta$, i.e. given ϵ, a *fixed* value of δ can be selected which will satisfy the alternative definition of continuity at *every point* in the domain of f.

A mapping f of a metric space E into a metric space F is said to be a **homeomorphism** if it is a bijection and if it and its inverse $f^{-1}: F \rightarrow E$ are both continuous. Two spaces E and F are said to be **homeomorphic** if a homeomorphism exists mapping E to F.

Example. Let E_1 and E_2 be $\{x \in \mathbb{R} \mid 0 < x < 1\}$ and $\mathbb{R}$, respectively, each with the Euclidean metric and let $f: E_1 \to E_2$ be defined by

$$x \mapsto \frac{1}{1-x} - \frac{1}{x}.$$

Then f is *continuous* at every point of E_1, i.e. it is *continuous* in E_1. f is a bijection and its inverse $f^{-1}: E_2 \to E_1$, defined by

$$x \mapsto \begin{cases} \frac{1}{2} - \frac{1}{x} - \sqrt{\left(\frac{1}{x^2} + \frac{1}{4}\right)} & (x < 0), \\ \frac{1}{2} & (x = 0), \\ \frac{1}{2} - \frac{1}{x} + \sqrt{\left(\frac{1}{x^2} + \frac{1}{4}\right)} & (x > 0) \end{cases}$$

is *continuous* in E_2. It follows that f is a *homeomorphism* and that E_1 and E_2 are *homeomorphic*. An example of a *uniformly continuous function* is given on p. 119.

21 Topological spaces and continuity

The definition of continuity given in the previous section was based upon the notion of an open set which in its turn hung on the definition of a metric. In this section we extend the idea of continuity to functions defined on more general spaces by relaxing our demand that E should possess a metric and by basing our definitions on the concept of 'openness' which we shall *endow* with certain properties which in metric spaces *result* from our definition of an open set.

A **topological space** is a set E together with a family, $\mathscr{U}$, of subsets of E (i.e. $\mathscr{U} \subset \mathscr{P}(E)$) known as **open sets** in E, satisfying:

(*a*) $\varnothing \in \mathscr{U}$, $E \in \mathscr{U}$;

(*b*) if $U_1, \ldots, U_n \in \mathscr{U}$ then $\bigcap_{i=1}^{n} U_i \in \mathscr{U}$;

(*c*) if $\mathscr{V} \subset \mathscr{U}$ then $\bigcup_{U \in \mathscr{V}} U \in \mathscr{U}$ (p. 10).

We say that $\mathscr{U}$ is a **topology** on E.

Note that the intersection of a *finite* number and the union of *any* number of elements of $\mathscr{U}$ must belong to $\mathscr{U}$.

A set $Y \subset E$ is **closed** if its complement, $E - Y$, is an open set.

Just as the definition of a closed set in a topological space takes the same form as that of a closed set in a metric space, so do the definitions of **open neighbourhood** (p. 101), **interior** (p. 101), **cluster** and **limit points** (p. 101), **closure** (p. 102), **frontier** (p. 102), **dense** (p. 102) and **separable** (p. 103).

A set $\mathscr{B} \subset \mathscr{P}(E)$ is said to be a **basis** for a topology on E if:

(*a*) $\varnothing \in \mathscr{B}$;

(*b*) $\bigcup_{B \in \mathscr{B}} B = E$;

(*c*) for all $B_1, B_2 \in \mathscr{B}$, there exists $\mathscr{B}' \subset \mathscr{B}$ such that

$$B_1 \cap B_2 = \bigcup_{B \in \mathscr{B}'} B.$$

Setting $\mathscr{U}_{\mathscr{B}} = \{U \in \mathscr{P}(E) | U \text{ is a union of elements of } \mathscr{B}\}$, then $(E, \mathscr{U}_{\mathscr{B}})$ is a topological space. $\mathscr{U}_{\mathscr{B}}$ is the **topology** on E **generated by** $\mathscr{B}$.

Example. The set of open balls (p. 100) forms a basis for a topology on a metric space E. Indeed a metric space has the structure of a topological space in which the open sets are unions of balls. Thus, given a metric space (E, d) we can ignore the metric and consider only E and its open and closed sets. E is then a topological space which is **metrisable** by d.

If one takes $\mathscr{U}$, the family of open sets of E, to be $\mathscr{P}(E)$, then one obtains the *discrete topology* on E, which is *metrisable* by the *discrete metric* (p. 99). If one takes $\mathscr{V} = \{\varnothing, E\}$, then one obtains the *indiscrete topology* on E which, provided E contains at least two points, is *non-metrisable*.

Two bases, $\mathscr{B}_1$ and $\mathscr{B}_2$, are said to be **equivalent** if they generate the same topology, i.e. if $\mathscr{U}_{\mathscr{B}_1} = \mathscr{U}_{\mathscr{B}_2}$.

It follows that two different metrics on a set E can give rise to equivalent topologies although they need not necessarily do so (thus when a topological space is metrisable, the metric, d, is not uniquely determined).

If $\mathscr{U}$ is a topology on E and $X \subset E$, then

$$\mathscr{U}_X = \{U \cap X \mid U \in \mathscr{U}\}$$

is a topology on X known as the **induced** or **relative topology.**

If E_1 and E_2 are topological spaces and f a function $E_1 \rightarrow E_2$, then we say that f is **continuous** if the inverse images (p. 15) of open sets are open, i.e., f is continuous if, for all open sets $V \subset E_2$, $f^{-1}(V)$ is open in E_1.

In the case of spaces which are metrisable, this definition is equivalent to those given on p. 103.

As in the case of metric spaces, we say that $f: (E, \mathscr{U}) \rightarrow (F, \mathscr{V})$ is a **homeomorphism** if it is a bijection and if both f and f^{-1} are continuous. E and F are said to be **homeomorphic** if there exists a homeomorphism mapping E to F.

Let $f: E \rightarrow F$ be a surjection from the topological space E onto the *set* F. The **identification topology** on F determined by f – the **identification map** – is that in which $Y \subset F$ is closed if and only if $f^{-1}(Y)$ is closed. (f is then a continuous function.)

If $\mathscr{U}$ and $\mathscr{V}$ are two topologies defined on the set E, we say that $\mathscr{U}$ is **weaker** than $\mathscr{V}$ (or $\mathscr{U}$ is **coarser** than $\mathscr{V}$) if $\mathscr{U} \subset \mathscr{V}$ (or, what is equivalent, if the identity map $id: (E, \mathscr{V}) \rightarrow (E, \mathscr{U})$ is continuous). $\mathscr{V}$ is then **finer** than, or a **refinement of,** $\mathscr{U}$.

Example. The indiscrete topology (see above) is the *coarsest* topology on E, and the discrete topology is *finer than* any other topology on E.

Given two spaces $(E, \mathscr{U})$ and $(F, \mathscr{V})$ we define their **topological product** $E \times F$ to be the set of pairs $\{(x, y) \mid x \in E, y \in F\}$ with the topology in which a basis of open sets consists of the sets $U \times V$ where $U \in \mathscr{U}$ and $V \in \mathscr{V}$.

Let $(E, \mathcal{U})$ be a topological space. We say that:

(*a*) E is a T_0 space if, for all distinct $x_1, x_2 \in E$, there exists $U \in \mathcal{U}$ such that *either* $x_1 \in U$, $x_2 \notin U$ *or* $x_2 \in U$, $x_1 \notin U$;

(*b*) E is a T_1 space if, for all distinct $x_1, x_2 \in E$, there exist $U_1, U_2 \in \mathcal{U}$ such that $x_1 \in U_1$, $x_2 \in U_2$, $x_1 \notin U_2$, $x_2 \notin U_1$;

(*c*) E is a T_2 space or a **Hausdorff space** if, for all distinct $x_1, x_2 \in E$, there exist $U_1, U_2 \in \mathcal{U}$ such that $x_1 \in U_1$, $x_2 \in U_2$ and $U_1 \cap U_2 = \varnothing$;

(*d*) E is **regular** or T_3 if E is T_1 and for all closed sets X in E and each $x \in E - X$, there exist disjoint open sets U_1 and U_2 such that $x \in U_1$ and $X \subset U_2$.

(*e*) E is **normal** or T_4 if E is T_1 and for every pair X_1, X_2 of disjoint closed sets in E there exist disjoint open sets U_1, U_2 such that $X_i \subset U_i$ $(i = 1, 2)$.

Note. (i) Every metric space is normal and, *a fortiori*, a Hausdorff space.

(ii) An equivalent definition of a Hausdorff space is that any two distinct points of the set should possess disjoint neighbourhoods.

(iii) Any $T_k (k = 0, 1, \ldots, 4)$ space is a T_i space for $i < k$.

In any topological space E we say that B is a **Borel set** if it can be obtained by a countable number of operations, starting from open sets, each operation consisting of forming unions, intersections or complements. Similarly, the sets obtained starting from compact sets (p. 110) are known as **Borelian sets**.

The classes of Borel and Borelian sets form σ-fields (p. 79).

Example. Let $E = \{x_1, x_2\}$ and $\mathcal{W} = \{\varnothing, E, \{x_1\}\}$. Then the topological space $(E, \mathcal{W})$ is a T_0 space but *not* a T_1 space, whereas $(E, \mathcal{U})$ (p. 106) is trivially *normal* (T_4), for we need only take $U_i = X_i$. It is easy to construct *Borel sets* in any given topological space, on p. 187, however, we give an example of a set in $\mathbb{R}$ which is *not* a Borel set.

In the following section we shall revert to considering metric spaces. Some of the results and notions will, however, be applicable to more general topological spaces. Such ideas as, for example, neighbourhoods, closed sets, limit points, closure, dense and perfect sets, continuous functions and homeomorphism are by nature topological. Others such as ball, sphere, diameter, distance from, bounded set and uniformly continuous functions are not topological notions and apply only to metric spaces.

22 Metric spaces II

Limits and sequences

Let (E, d_1) and (F, d_2) be metric spaces, X a subset of E and f a function $X \to F$. We say that $f(x)$ has the **limit** $l \in F$ **as** $x \in X$ **tends to** $a \in X'$ if, for every neighbourhood V of l in F, there exists a neighbourhood U of a in E such that

$$f((U - \{a\}) \cap X) \subset V.$$

We then write $l = \lim_{x \to a,\, x \in X} f(x)$, or $f(x) \to l$ as $x \to a$ $(x \in X)$.

Analogous to the $\epsilon - \delta$ definition of continuity (p. 103), we have the alternative definition:

l is the **limit** of $f(x)$ when $x (\in X)$ tends to a if and only if $\forall \epsilon > 0, \exists \delta > 0$ such that $d_2(f(x), l) < \epsilon$ whenever $0 < d_1(x, a) < \delta (x \in X)$.

Note. (i) If $a \in X$, then f is *continuous* at a if and only if

$$\lim_{x \to a,\, x \in X} f(x) = f(a).$$

(ii) Often we have $E = X$ in which case we write $l = \lim_{x \to a} f(x)$.

Functions of particular interest are those which map the natural numbers $\mathbb{N}$ into a metric space E. Such a function is called a **sequence** of points of E and is denoted by $n \mapsto x_n$, (x_n) or $(x_0, x_1, x_2, \ldots)$.

Note. (i) A common alternative notation for the sequence (x_n) is $\{x_n\}$. This latter notation does, however, tend to obscure the fundamental difference between a function and a set and can lead to confusion in the case of sequences such as $(1, 1, 1, \ldots)$.

(ii) This definition of a sequence is confusing in that it suggests that one might wish to make use of, say, the semigroup structure of $\mathbb{N}$ in subsequent theory whereas the significant feature is that the domain of the function is a well-ordered set of order type ω (p. 80).

Given $(x_n) : \mathbb{N} \to E$, we say that l is the **limit of the sequence** (x_n) (or that the sequence **converges to** or has the **limit** l) if for all neighbourhoods V of l there exists an integer N such that $n > N$ implies $x_n \in V$, i.e. if every neighbourhood of l contains all but a finite number of points of the sequence.

Alternatively, (x_n) converges to l if and only if for all $\epsilon > 0$ there exists an integer N such that $d(l, x_n) < \epsilon$ whenever $n > N$.

A sequence which possesses a limit is said to be **convergent.** If (x_n) converges to the limit l, we write

$$\lim_{n\to\infty} x_n = l \quad \text{or} \quad x_n \to l \quad \text{as} \quad n \to \infty.$$

If the limit of a sequence is defined before that of a function, one can give an alternative definition of the latter as follows:

Let E and F be metric spaces, X a subset of E and f a function $f: X \to F$. We say that f has a **limit** $l \in F$ when x tends to $a \in \bar{X}$ if and only if for every sequence (x_n) of points of X such that $a = \lim_{n\to\infty} x_n$, we have $\lim_{n\to\infty} f(x_n) = l$.

We can, similarly, base our definition of continuity on that of the limit of a sequence by saying:

$f: X \to F$ is **continuous** at $a \in X$ if and only if for every sequence (x_n) converging to a, the sequence $(f(x_n))$ converges to $f(a)$, i.e. if

$$f(\lim_{n\to\infty} x_n) = \lim_{n\to\infty} f(x_n).$$

A **subsequence** of $f: \mathbb{N} \to E$ is a composite function $f \circ g: \mathbb{N} \to E$, where $g: \mathbb{N} \to \mathbb{N}$ is an *increasing sequence*, i.e. $g(n) < g(m)$ if $n < m$.

This definition formalises the notion of a subsequence as a sequence derived from the original sequence by the omission of a number (not necessarily finite) of terms.

Examples. The terms introduced so far in this section are considered again in § 24 and examples can be found there.

Given a metric space (E, d) we define a **Cauchy sequence** (or **fundamental sequence**) to be a sequence (x_n) for which, given any $\epsilon > 0$, there exists $N \in \mathbb{N}$ such that

$$d(x_m, x_n) < \epsilon \quad \text{for all} \quad m, n \geqslant N.$$

It follows from our previous definition that any convergent sequence is a Cauchy sequence.

A metric space (E, d) is said to be **complete** if any Cauchy sequence in E is convergent, i.e. if given any Cauchy sequence (x_n) in E there is a point $l \in E$ such that $l = \lim x_n$.

Example. The metric space $\mathbb{Q}$ (the set of rationals with the Pythagorean metric) is not complete since it contains Cauchy sequences which are not convergent. However, these same sequences considered as sequences of points of $\mathbb{R}$ do converge (i.e. $\mathbb{R}$ is complete). These facts form the basis of the construction of $\mathbb{R}$ given in the next section.

Compactness and connectedness

A collection, $\mathscr{C}$, of subsets of E is said to **cover** $X \subset E$ (or to form a **covering** for X) if $X \subset \bigcup_{C \in \mathscr{C}} C$. If all the sets of $\mathscr{C}$ are open, $\mathscr{C}$ is called an **open covering.**

A subset X of E is said to be **compact** if, for each open covering $\mathscr{C}$ of X, there is a finite subset $\mathscr{C}_1 \subset \mathscr{C}$ such that $\mathscr{C}_1$ covers X.

A subset X of (E, d) is **precompact** if, for any given $\epsilon > 0$, there is a finite covering of X by sets of diameter less than ϵ.

Note that *compact* is a topological notion whereas *precompact* is metric.

A space E is said to be **locally compact** if every point of E possesses a neighbourhood the closure of which is compact.

If $A \subset E$ is such that $\bar{A}$ is compact, we say that A is **relatively compact.**

A space E is said to be **sequentially compact** if every sequence of points of E has a convergent subsequence. (Since, in metric spaces, sequential compactness is equivalent to compactness this yields an alternative definition of a compact metric space.)

A collection $\mathscr{A}$ of subsets of E is said to have the **finite intersection property** (f.i.p.) if every finite intersection

$$A_1 \cap A_2 \cap \ldots \cap A_n \quad (A_i \in \varnothing)$$

is non-empty.

A topological space E is said to have the **finite intersection property** if every collection $\mathscr{A}$ of closed sets having the f.i.p. also satisfies $\bigcap_{A \in \mathscr{A}} A \neq \varnothing$.

Note. A topological space E is compact if and only if it has the finite intersection property.

A space E is said to be **connected** if E has no proper subsets which are both open and closed.

Alternatively, E is **connected** if and only if it is not the union of two disjoint, non-empty, open sets.

A subset $X \subset E$ is **connected** if it is connected in the relative topology (p. 106).

E is **locally connected** if, for every $x \in E$, there is a fundamental system (p. 101) of connected neighbourhoods of x.

Given $x \in X \subset E$, then the union $C(x)$ of all connected subsets of X which contain x is called the **connected component** of x in X. ($C(x)$ is the largest connected set in X containing x.) If every connected component of X is a singleton, we say that X is **totally disconnected**.

Examples. (i) The set $X = \{x \in \mathbb{R} \mid a \leqslant x \leqslant b\}$ is *compact* and also *precompact*. $\mathbb{R}$ is *not* compact and, since continuous mappings preserve compactness, it follows (see example p. 104) that $\{x \in \mathbb{R} \mid 0 < x < 1\}$ is *not* compact. $\mathbb{R}$ is, however, *locally compact*. Any bounded subset (p. 100) of $\mathbb{R}$ is *relatively compact* (see Heine–Borel Theorem p. 204).

(ii) Any ball in $\mathbb{R}^n$ is *connected*. $\mathbb{R} - \{0\}$ is *not* connected, for, in the induced (relative) topology, the sets $\{x \in \mathbb{R} \mid x > 0\}$ and $\{x \in \mathbb{R} \mid x < 0\}$ are both open and closed. The *connected component*, $C(2)$, of 2 in $\mathbb{R} - \{0\}$ is $\{x \in \mathbb{R} \mid x > 0\}$. $\mathbb{R}^n - \{0\}$ is *connected* for $n > 1$. See also the example on p. 152.

Normed spaces

On p. 90 we showed how on an inner-product space we can define a *norm* which (see p. 99) gives rise to an obvious metric.

In general, if E is a vector space over $\mathbb{R}$ (or $\mathbb{C}$), then we define a **norm** on E to be a mapping $E \to \mathbb{R}$, denoted by $x \mapsto \|x\|$, which satisfies:

(*a*) $\|x\| \geqslant 0$ for all $x \in E$,

(*b*) $\|x\| = 0 \Leftrightarrow x = 0$,

(*c*) $\|\lambda x\| = |\lambda| \, \|x\|$ for all $x \in E$ and $\lambda \in \mathbb{R}$,

(*d*) $\|x+y\| \leqslant \|x\| + \|y\|$ for all $x, y \in E$.

A space with a norm is known as a **normed space**; it is naturally (p. 99) a metric space with metric $d(x, y) = \|x-y\|$.

A normed space which is complete (p. 109) is known as a **Banach space**.

As with metrics, two norms are said to be *equivalent*, if they generate the same topology on E.

If E and F are two normed spaces, then it can be shown that the set of all continuous linear mappings of E into F, denoted by $\mathscr{L}(E, F)$ is a vector space. If E and F are Banach spaces, then so is $\mathscr{L}(E, F)$ when the latter is 'normed' as in the examples below.

Examples. (i) $\mathbb{R}^2$ together with the norm $(x_1, x_2) \mapsto |x_1| + |x_2|$ is a *normed space* which is complete and hence a *Banach space*. This norm is *equivalent* to the standard Euclidean norm $(x_1, x_2) \mapsto \sqrt{(x_1^2 + x_2^2)}$.

(ii) The vector space $\mathscr{L}(E, F)$ can be *normed* in the following way. Given $u \in \mathscr{L}(E, F)$, we define $\|u\|$ to be the greatest lower bound of all the constants

$a > 0$ which satisfy $\|u(x)\| \leqslant a \,.\, \|x\|$ for all $x \in E$. Equivalently, we may define $\|u\| = \operatorname{lub}_{\|x\| \leqslant 1} \|u(x)\|$.

In the special case where the norm arises from an inner product (p. 90), we say that the space is **prehilbert.**

Prehilbert spaces encompass vector spaces over $\mathbb{C}$ on which is defined a positive bilinear form satisfying $f(u, v) = \overline{f(v, u)}$ (such forms are also known as **positive Hermitian forms**) (see note on p. 92).

A **Hilbert space** is a prehilbert space which is complete.

All Hilbert spaces are, *a fortiori*, Banach spaces, indeed any Hilbert space of dimension n over $\mathbb{R}$ is isomorphic to $\mathbb{R}^n$ with the usual scalar (inner) product.

A Hilbert space of particular importance is the space denoted by H^∞ or ℓ_2 whose points are infinite sequences of real numbers (u_n) such that $\sum_{n=0}^{\infty} u_n^2$ (p. 140) converges and which has the inner product

$$\langle (u_n), (v_n) \rangle = \sum_{n=0}^{\infty} u_n v_n.$$

Note. Some definitions of a Hilbert space demand that the space should be separable (as is H^∞).

See also: Heine–Borel Theorem (p. 204), Heine's Theorem (p. 204).

23 The real numbers

We begin the second method of constructing the set $\mathbb{R}$ by considering the set $\mathbb{Q}^{\mathbb{N}}$ of all sequences $f: \mathbb{N} \to \mathbb{Q}$. If addition and multiplication of sequences are defined by $(x_n)+(y_n)=(x_n+y_n)$ and $(x_n)(y_n)=(x_n y_n)$ respectively, then the subset of $\mathbb{Q}^{\mathbb{N}}$ consisting of all Cauchy sequences (p. 109) forms a ring which we denote by $\mathscr{S}_C(\mathbb{Q})$. Moreover, the set of all Cauchy sequences which converge to $0 \in \mathbb{Q}$, a set denoted by $\mathscr{S}_0(\mathbb{Q})$, is an ideal (p. 32) of $\mathscr{S}_C(\mathbb{Q})$.

$\mathbb{R}$, the **set of real numbers**, is defined to be the quotient algebra (p. 36)

$$\mathscr{S}_C(\mathbb{Q})/\mathscr{S}_0(\mathbb{Q}).$$

It is a consequence of this definition that, under the induced operations of addition and multiplication, $\mathbb{R}$ is a commutative ring. Furthermore, given $a \in \mathbb{Q}$ we can define a map $p: \mathbb{Q} \to \mathbb{R}$ by defining $p(a)$ to be the residue class containing the constant sequence $(a, a, a, \ldots)$. The mapping p preserves addition and multiplication, i.e. it is a ring-monomorphism of $\mathbb{Q}$ into $\mathbb{R}$, and we can therefore identify $\mathbb{Q}$ with a subring of $\mathbb{R}$.

It can be further shown that $\mathbb{R}$ is not only a ring but a field.

An order relation is defined upon $\mathbb{R}$ by:

Given $\alpha, \beta \in \mathbb{R}$, then $\alpha \leqslant \beta$ if there exist (a_n) in the coset α and (b_n) in the coset β such that $a_n \leqslant b_n$ for *sufficiently large n*, i.e. there exists $N \in \mathbb{N}$ such that $a_n \leqslant b_n$ for all $n > N$.

The relation $\leqslant$ is then a total ordering (p. 75) of $\mathbb{R}$, and $(\mathbb{R}, \leqslant)$ is an ordered field.

If we now attempt to repeat this construction by considering the subring $\mathscr{S}_C(\mathbb{R})$ of $\mathbb{R}^{\mathbb{N}}$ and forming the quotient algebra

$$\mathscr{S}_C(\mathbb{R})/\mathscr{S}_0(\mathbb{R}),$$

we find that the resulting algebra is isomorphic to $\mathbb{R}$. This happens since $\mathbb{R}$ is complete (p. 109) and all Cauchy sequences converge in $\mathbb{R}$.

$\mathbb{R}$ is, therefore, a complete ordered field; indeed, it is the only ordered field which is complete.

Note. This construction of $\mathbb{R}$ is essentially due to Cantor and for that reason the elements of $\mathscr{S}_C(\mathbb{Q})$ are often referred to as **Cantor fundamental sequences.**

Decimals

Consider $\alpha \in \mathbb{R}$ satisfying $\alpha > 0$. Since $\alpha > 0$ there must be an $a_0 \in \mathbb{N}$ such that $a_0 \leqslant \alpha < a_0 + 1$. Consider $\alpha_1 = 10(\alpha - a_0)$. We then have $0 \leqslant \alpha_1 < 10$ and so can find $a_1 \in \mathbb{N}$ such that (*a*) $0 \leqslant a_1 \leqslant 9$, (*b*) $a_1 \leqslant \alpha_1 < a_1 + 1$. Taking $\alpha_2 = 10(\alpha_1 - a_1)$ and repeating the process we obtain a sequence $(a_0, a_1, a_2, \ldots) \in \mathbb{N}^{\mathbb{N}}$ containing the subsequence $(a_1, a_2, \ldots) \in \{0, 1, 2, \ldots, 9\}^{\mathbb{N}}$.

From this sequence we construct a second sequence

$$(a_0, a_0.a_1, a_0.a_1a_2, a_0.a_1a_2a_3, \ldots, a_0.a_1a_2\ldots a_n, \ldots),$$

where $a_0.a_1a_2 \ldots a_n$ denotes the rational number

$$q = a_0 + 10^{-1}a_1 + 10^{-2}a_2 + \ldots + 10^{-n}a_n$$

and is known as the **decimal representation** of q.

It can be readily checked that this second sequence is a Cauchy sequence, i.e. it is an element of $\mathscr{S}_C(\mathbb{Q})$, and that in the quotient algebra $\mathscr{S}_C(\mathbb{Q})/\mathscr{S}_0(\mathbb{Q})$ it is mapped onto α. Thus every positive real number can be represented by a Cauchy sequence of type

$$(a_0, a_0.a_1, a_0.a_1a_2, \ldots, a_0.a_1a_2 \ldots a_n, \ldots)$$

where $a_i \in \mathbb{N}$ (all i), $0 \leqslant a_i \leqslant 9$ $(i = 1, 2, 3, \ldots)$.

Such a representation is called a **decimal representation** for α. It is usual to abbreviate it to $a_0.a_1a_2 \ldots$. (The representation can, of course, be extended to cover negative real numbers.)

This representation can be shown to be unique unless α is a rational number which can be expressed in the form p/q with q a power of 10, e.g. $\frac{1}{2} = 5/10$. In this latter case two representations are possible, one of the form

$$a_0.a_1a_2\ldots a_{N-1}a_N000\ldots(a_N \neq 0)$$

and the other of the form

$$a_0.a_1a_2\ldots a_{N-1}(a_N - 1)\,999\ldots.$$

The first form is generally written $a_0.a_1a_2\ldots a_N$ and the decimal is said to **terminate.**

A representation of the type

$$a_0.a_1a_2\ldots a_{p-1}a_p\ldots a_qa_p\ldots a_qa_p\ldots a_qa_p\ldots,$$

in which one group of digits, $a_p\ldots a_q$, recurs, is known as a **recurring decimal** (e.g. 2.3786868686...).

It is easily shown that α is represented by a recurring or terminating decimal if and only if it is rational.

24 Real-valued functions of a real variable

Because of the importance of functions whose domain and codomain are subsets of $\mathbb{R}$ we repeat certain key definitions presenting them in a less general and more elementary form.

Given a sequence (x_n) of real numbers, we say that it is:
(*a*) **bounded above** if there exists $U \in \mathbb{R}$ such that $x_n \leqslant U$ for all n,
(*b*) **bounded below** if there exists $L \in \mathbb{R}$ such that $x_n \geqslant L$ for all n,
(*c*) **bounded** if it is bounded above and below,
(*d*) **convergent** if it has a *limit*, l, i.e. if there exists $l \in \mathbb{R}$ for which, given any $\epsilon > 0$ there exists $N \in \mathbb{N}$ (N depending on ϵ) such that $|x_n - l| < \epsilon$ for all $n > N$ (we then write '$\lim_{n\to\infty} x_n = l$' or '$x_n \to l$ as $n \to \infty$'),

It is a consequence of the properties of $\mathbb{R}$ (see p. 113) that a sequence (x_n) of real numbers is convergent in $\mathbb{R}$ if and only if it is a *Cauchy sequence* (p. 109), i.e. if, for all $\epsilon > 0$, there exists an $N \in \mathbb{N}$ such that $|x_p - x_q| < \epsilon$ for all integers p and q both greater than N. This condition is known as the **Cauchy condition for the convergence of sequences** or as the **General Principle of Convergence.**

(*e*) **divergent** if it is not convergent,
(*f*) **properly divergent** if, given any $A \in \mathbb{R}$, there exists $N \in \mathbb{N}$ (N depending on A) such that $x_n > A$ for all $n > N$ (in which case we write $x_n \to +\infty$) or such that $x_n < A$ for all $n > N$ (in which case we write $x_n \to -\infty$),
(*g*) **oscillating** if it is divergent but not properly divergent – if bounded the sequence is said to **oscillate finitely**, if unbounded to **oscillate infinitely**,
(*h*) **monotonic increasing** if $x_{n+1} \geqslant x_n$ for all n,
(*i*) **monotonic decreasing** if $x_{n+1} \leqslant x_n$ for all n,
(*j*) **strictly monotonic** if the inequality in (*h*) or (*i*) is *strict* (i.e. is $>$ or $<$).

Examples. (i) The *sequence* $1, -1, 1, -1, 1, -1, \ldots$ is *bounded* since it is *bounded above* by 2, say, and *bounded below* by -1. It is *not* convergent, since if we take $\epsilon = \frac{1}{2}$ then there is no l and no N for which $|x_n - l| < \frac{1}{2}$ for all $n > N$. It is *not* properly divergent, since there is no N such that $x_n > 2$ for all $n > N$ or such that $x_n < -1$ for all $n > N$. The sequence *oscillates finitely*.

(ii) The sequence $(1 - 1/(n+1)^2)$ (i.e. the sequence $0, \frac{3}{4}, \frac{8}{9}, \frac{15}{16}, \ldots$) is *strictly monotonic increasing* and tends to the *limit* 1. It is, therefore, *convergent*. We note that, given any $\epsilon > 0$, then if we take $N > \epsilon^{-\frac{1}{2}}$ we have $|x_n - 1| < \epsilon$ whenever $n > N$.

Subsets of $\mathbb{R}$ which frequently occur as domains of functions include:

(*a*) **open intervals**, i.e. sets of the type $\{x \in \mathbb{R} | a < x < b\}$, which are denoted by $< a, b >$ (or $]a, b[$ or (a, b));

(*b*) **closed intervals**, i.e. sets of the type $\{x \in \mathbb{R} | a \leqslant x \leqslant b\}$, denoted by $\leqslant a, b \geqslant$ (or $[a, b]$ or $\langle a, b \rangle$);

(*c*) **half-open intervals**, i.e. sets of the type $\{x \in \mathbb{R} | a \leqslant x < b\}$ and $\{x \in \mathbb{R} | a < x \leqslant b\}$, denoted by $\leqslant a, b >$ and $< a, b \geqslant$ respectively.

Such intervals are termed *finite*. Sometimes we wish to consider *infinite* or *unbounded intervals* such as $\{x \in \mathbb{R} | x < a\}$, $\{x \in \mathbb{R} | x \leqslant a\}$, $\{x \in \mathbb{R} | x > a\}$ and $\{x \in \mathbb{R} | x \geqslant a\}$ which will be denoted by $< -\infty, a >$, $< -\infty, a \geqslant$, $< a, \infty >$ and $\leqslant a, \infty >$, respectively.

a and b are called the **end-points** of the interval and $|b-a|$ is the **length** of the interval.

Note. (i) An open interval is an open set (p. 101) of the metric space obtained by taking the Euclidean metric (p. 99) on $\mathbb{R}$.

(ii) If I is any interval of $\mathbb{R}$, then the set $\mathbb{R}^I$ (p. 13) of functions with domain I and codomain $\mathbb{R}$, can be shown to be a *commutative algebra* over $\mathbb{R}$ (p. 39) if addition and multiplication are defined by:

$$(f+g)(x) = f(x)+g(x),$$
$$(f.g)(x) = f(x).g(x),$$
$$(\lambda f)(x) = \lambda f(x),$$

for all $f, g \in \mathbb{R}^I$ and $\lambda \in \mathbb{R}$.

The function $f.g$ is known as the **product** of the functions f and g and, as pointed out on p. 15, it must not be confused with the *composite* of f and g.

Let I be an interval of $\mathbb{R}$ and f a function $I \to \mathbb{R}$. We say that f is **continuous** at $a \in I$ if for every sequence (x_n) converging to a, the sequence $(f(x_n))$ converges to $f(a)$, i.e. if $f(\lim x_n) = \lim f(x_n)$.

f is said to be **continuous** on I if it is continuous at every $a \in I$.

An alternative sequence of definitions is:

The function $f: I \to \mathbb{R}$ is said to have the **limit** l as x tends to $a \in \bar{I}$ (written '$\lim_{x \to a} f(x) = l$' or '$f(x) \to l$ as $x \to a$') if, given any $\epsilon > 0$, there exists a δ (depending upon ϵ) such that $|f(x) - l| < \epsilon$ for all $x \in I$ satisfying $0 < |x-a| < \delta$. (We then also say that f **converges** to l as x **converges** to a.)

$f: I \to \mathbb{R}$ is said to be **continuous** at $a \in I$ if $\lim_{x \to a} f(x)$ exists and is equal to $f(a)$.

Equivalently, f is continuous at $a \in I$ if, given any $\epsilon > 0$, there exists $\delta > 0$ such that $|f(x) - f(a)| < \epsilon$ for all $x \in I$ satisfying $|x - a| < \delta$.

Again, f is **continuous** on I if it is continuous at every point of I.

Note. The set of all continuous functions defined on an interval I forms a subalgebra (p. 39), $\mathscr{C}(I)$, of the algebra $\mathbb{R}^I$ (p. 116), since it is readily shown that sums, products and scalar multiples of continuous functions are continuous.

$f: I \to \mathbb{R}$ is said to **tend to infinity** as $x \to a$ (written '$\lim_{x \to a} f(x) = \infty$' or '$f(x) \to \infty$ as $x \to a$') if, given any $\Delta \in \mathbb{R}$, there exists $\delta > 0$ such that $f(x) > \Delta$ for all $x \in I$ satisfying $0 < |x - a| < \delta$.

We write $\lim_{x \to a} f(x) = -\infty$ if $\lim_{x \to a} (-f(x)) = \infty$.

Example. (i) The function

$$f: \begin{cases} x \mapsto \sin 1/x & (x \neq 0), \\ x \mapsto 0 & (x = 0) \end{cases}$$

is *not* continuous at $x = 0$, since if we set $a_n = 2/(4n+1)\pi$, $b_n = 2/(4n+3)\pi$. Then, when $n \to \infty$, $a_n \to 0$, $b_n \to 0$, $f(a_n) \to 1$, $f(b_n) \to -1$, i.e.,

$$\lim f(a_n) \neq \lim f(b_n) \neq f(0).$$

(ii) The function

$$g: \begin{cases} x \mapsto x \sin 1/x & (x \neq 0), \\ x \mapsto 0 & (x = 0) \end{cases}$$

is *continuous* on $\mathbb{R}$ and, in particular, at $x = 0$. For given $\epsilon > 0$, and taking $\delta = \epsilon$, we have $|g(x) - 0| = |x \sin 1/x| \leqslant |x| < \epsilon$ provided $0 < |x| < \delta$.

(iii) The function

$$h: \begin{cases} x \mapsto x \sin 1/x & (x \neq 0), \\ x \mapsto 1 & (x = 0) \end{cases}$$

is *not* continuous at $x = 0$, since $\lim_{x \to 0} h(x) = 0 \neq h(0)$.

$f: I \to \mathbb{R}$ is said to have a **limit** l as x tends to $a \in \bar{I}$ **from the right (from above)** if, given any $\epsilon > 0$, there exists a $\delta > 0$ such that $|f(x) - l| < \epsilon$ for all $x \in I$ satisfying $a < x < a + \delta$. We then write $\lim_{x \to a+} f(x) = l$. The limit is undefined if $I \cap \langle a, a+\delta \rangle = \phi$.)

We can define a **limit from the left** (or **from below**) similarly.

f is **continuous on the right** at $a \in I$ if $\lim_{x \to a+} f(x) = f(a)$, and is **continuous on the left** if $\lim_{x \to a-} f(x) = f(a)$.

f is a **regulated function** if it possesses limits from the left and from the right (not necessarily equal) at all points of $\mathscr{I}(I)$ (p. 101), and the appropriate limits at points of $\mathscr{F}(I)$ (p. 102). Important examples of the latter are so-called **step functions**, i.e. functions defined on an interval $I = \langle a, b \rangle$ in such a way that they are constant on each open subinterval $< a_i, a_{i+1} >$ $(i = 0, 1, \ldots, n-1)$ where a_i $(i = 0, 1, \ldots, n)$ is some increasing sequence of points of I such that $a_0 = a$, $a_n = b$. A finite set of points $P = \{a_0, \ldots, a_n\}$ satisfying the above requirements, i.e.

$$a = a_0 < a_1 < a_2 < \ldots < a_n = b,$$

is known as a **partition of the interval** $\leqslant a, b \geqslant$. The interval $< a_{k-1}, a_k >$ is known as the kth interval and we write $\Delta a_k = a_k - a_{k-1}$ so that $\sum_{i=1}^{n} \Delta a_k = b - a$. The partition P' is said to be **finer** than, or a **refinement of**, P if $P \subset P'$. Given a partition P we define the **mesh** of P to be $\max \Delta a_i$ and we denote it by $\mu(P)$.

A function $f: \leqslant a, b \geqslant \; \to \mathbb{R}$ is said to be of **bounded variation** on $\leqslant a, b \geqslant$ if there exists a positive number M such that for every partition $P = \{a_0, a_1, \ldots, a_n\}$ of $\leqslant a, b \geqslant$

$$\sum_{k=1}^{n} |(f(a_k) - f(a_{k-1})| \leqslant M.$$

In particular, if f is monotonic (p. 119) on $\leqslant a, b \geqslant$, then it is of bounded variation. It can be shown that any function of bounded variation can be expressed as the difference of two monotonic increasing functions.

A function f defined on the closed interval $I = \; \leqslant a, b \geqslant$ is said to be **piecewise continuous** on I if:

(i) $\lim_{x \to a+} f(x)$ and $\lim_{x \to b-} f(x)$ both exist;

(ii) $f(x)$ is continuous at all but a finite number of points in $< a, b >$;

(iii) left and right limits exist at all points of $< a, b >$.

A function $f: I \to \mathbb{R}$ is said to be **uniformly continuous** on I if, in the $\epsilon - \delta$ definition of continuity (p. 117), given $\epsilon > 0$ we can choose a δ which depends upon ϵ but does not depend upon the particular value of $a \in I$ under consideration.

Examples. The function $f: x \mapsto 1/x$ with domain $\mathbb{R} - \{0\}$ *tends to infinity* as x tends to 0 from the right, i.e. $f(x) \to \infty$ as $x \to 0+$. We note that $f(x) \to -\infty$ as $x \to 0-$.

The **square bracket function, Legendre function**, or **greatest integer function** $x \mapsto [x]$ where $[x]$ is the largest integer which is not greater than x, is *continuous on the right* at all $a \in \mathbb{R}$. However, it is discontinuous on the left at all $n \in \mathbb{Z}$, for we have $\lim_{x \to n-} [x] = n - 1$, $[n] = n$. The square bracket function is *piecewise continuous* on any bounded closed interval of $\mathbb{R}$, for in any bounded interval it will only have a finite number of discontinuities and at every point of the interval left and right limits will exist. The square bracket function is a *step function* and, hence, a *regulated function* on $\mathbb{R}$. The function

g defined in the example on p. 117 is *regulated* in $\leqslant 0, 1 \geqslant$. However, if we define h by $x \mapsto 1$ $(x > 0)$, $0 \mapsto 0$, $x \mapsto -1$ $(x < 0)$, then, even though h is *regulated*, the composite function $h \circ g$ is *not*.

The function $x \mapsto 1/x$ is *continuous* but *not* uniformly continuous on $< 0, 1 \geqslant$, the function $x \mapsto x^2$ is *uniformly continuous* on $< 0, 1 \geqslant$. (In the latter case, given $\epsilon > 0$, then taking $\delta = \epsilon/2$ we can show that the function is continuous at any $a \in < 0, 1 \geqslant$.)

The function $x \mapsto x \sin 1/x$ $(x \neq 0)$, $0 \mapsto 0$, is *not* of bounded variation on $\leqslant 0, 2/\pi \geqslant$ although it is continuous on that interval. Consider the *partition*

$$P = \left\{0, \frac{2}{2n\pi}, \frac{2}{(2n-1)\pi}, \ldots, \frac{2}{3\pi}, \frac{2}{2\pi}, \frac{2}{\pi}\right\},$$

then

$$\sum_{k=1}^{2n} |f(a_k) - f(a_{k-1})|$$

$$= \left|\frac{2.1}{\pi} - 0\right| + \left|\frac{2}{2\pi}.0 - \frac{2}{3\pi}.(-1)\right| + \left|\frac{2}{3\pi}(-1) - 0\right| + \ldots + \left|\frac{\pm 2}{(2n-1)\pi} - 0\right|$$

$$= \frac{2}{\pi}\left[1 + \frac{2}{3} + \frac{2}{5} + \ldots + \frac{2}{2n-1}\right]$$

and this cannot be bounded since the series $1 + \frac{1}{3} + \frac{1}{5} + \ldots$ diverges.

A function $f: I \to \mathbb{R}$ is said to be:

(*a*) **monotonic increasing** if $f(u) \leqslant f(v)$ whenever $u \leqslant v$ in I,

(*b*) **monotonic decreasing** if $f(u) \geqslant f(v)$ whenever $u \leqslant v$ in I,

(*c*) **strictly monotonic** if the inequalities in (*a*) or (*b*) are strict (p. 115).

(*d*) **bounded** on I if there exists $M \in \mathbb{R}$ such that $|f(u)| < M$ for all $u \in I$.

A function $f: I \to \mathbb{R}$ is said to have a **(local) maximum** at $a \in I$ if there is a neighbourhood of a throughout which $f(x) < f(a)$ $(x \neq a)$, i.e. if there exists a δ such that for all $x \in I$ satisfying $0 < |x - a| < \delta$, $f(x) < f(a)$. If this condition is not satisfied but the weaker condition $f(x) \leqslant f(a)$ is, then f is said to have a **weak** or **improper maximum** at a.

The definitions of a **(local) minimum** and a **weak minimum** are obtained by reversing the inequality signs in an obvious way.

Example. The function $x \mapsto x^2$ is *strictly monotonic increasing* in the interval $\leqslant 0, \infty >$ and *strictly monotonic decreasing* in the interval $< -\infty, 0 \geqslant$. The function $\mathbb{R} \to \mathbb{R}$, $x \mapsto x^2$ has a *minimum* at 0. The square bracket function $[\,]: \mathbb{R} \to \mathbb{R}$ has a *weak minimum* and a *weak maximum* at $\frac{1}{2}$. It has a *weak maximum* at 1 but *not* a weak minimum there.

A function $f: I \to \mathbb{R}$, where I is an interval which is symmetrical about the origin (i.e., for all $a \in I$, $-a \in I$), is said to be

(i) **even** if for all $a \in I$, $f(a) = f(-a)$,

(ii) **odd** if for all $a \in I$, $f(a) = -f(-a)$.

Note. Any function defined on such an interval I can be expressed as the sum of an even and an odd function.

If f is a (non-constant) function from $\mathbb{R}$ into $\mathbb{R}$ and there exists a number q such that, for all $x \in \mathbb{R}$, $f(x+q) = f(x)$, then we say that f is a **periodic function.** The **period** of a periodic function is defined to be the smallest positive number p, if any, satisfying $\forall x \in \mathbb{R}, f(x+p) = f(x)$.

Example. The function $g: x \mapsto x \sin 1/x$ $(x \neq 0)$, $0 \mapsto 0$, is an *even function*, since $g(x) = g(-x)$, all $x \in \mathbb{R}$. The function $s: x \mapsto \sin 3x$ is an *odd function* which is *periodic* with *period* $2\pi/3$. The Dirichlet function (p. 202) is *periodic* since $\phi(x+q) = \phi(x)$ for all $q \in \mathbb{Q}$. However, it has no *period.*

Sequences of functions

Given a sequence (f_n), each term of which is a function with domain $I \subset \mathbb{R}$ and codomain $\mathbb{R}$, then corresponding to each $x \in I$ we can form another sequence $(f_n(x))$ whose terms are the values of f_n at x. Let T denote the subset of I consisting of those x for which the sequence $(f_n(x))$ converges (p. 115). The function f defined by the equation

$$f(x) = \lim_{n \to \infty} f_n(x) \quad \text{if} \quad x \in T$$

is known as the **limit function** of the sequence (f_n) and (f_n) is said to **converge pointwise** on the set T.

The conditions imposed by this definition mean that if (f_n) converges at the point $x \in T$, then, given any $\epsilon > 0$, there exists an $N \in \mathbb{N}$ (depending upon ϵ and x) such that $n > N$ implies

$$|f_n(x) - f(x)| < \epsilon.$$

As before (cf. p. 103 and p. 118), those instances in which one can select an N (or in other examples a δ) which depends solely on ϵ and *not* on the choice of x, are of particular importance. Again, this property is denoted by including the adverb *uniformly*, and we say, therefore that (f_n) **converges uniformly** to f on T, if, given any $\epsilon > 0$, we can find an $N \in \mathbb{N}$ such that the inequality above is satisfied for all $x \in T$. We then write

$$f_n \to f \textbf{ uniformly on } T.$$

A sequence (f_n) is said to be **pointwise bounded** on T if the sequence $(f_n(x))$ is bounded (p. 115) for every $x \in T$. (The bound will, in general, depend upon x.)

A sequence (f_n) is said to be **uniformly bounded** on T if in the previous definition a bound can be found which is valid for all x in T, i.e if there exists a number $M > 0$ such that $|f_n(x)| \leqslant M$ for all $x \in T$ and all $n \in \mathbb{N}$. M is then called a **uniform bound** for (f_n).

Examples. (i) Let $f_n(x) = x^n$, $I = \leqslant 0, 1 \geqslant$. Then (f_n) *converges pointwise* on I to the function $f: x \mapsto 0$ $(x \neq 1)$, $1 \mapsto 1$. The *non-uniformity* of the convergence of f on I is illustrated in the diagram, which also suggests (as is the case) that (f_n) *converges uniformly* on any closed interval $\leqslant 0, \delta \geqslant$ where $0 < \delta < 1$.

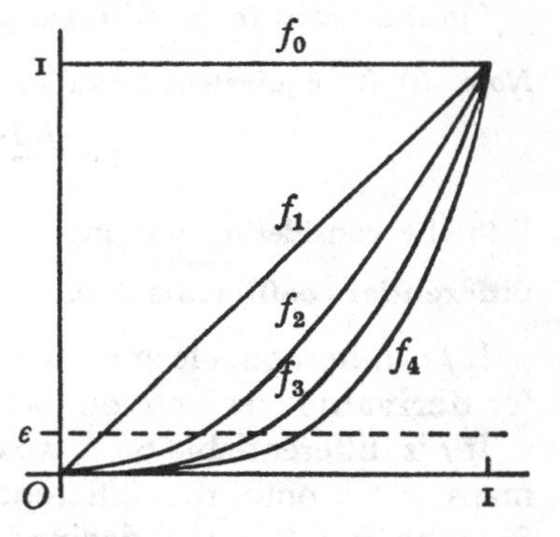

(ii) Let

$$f_n(x) = \frac{x^{2n}}{1+x^{2n}}, \quad x \in \mathbb{R}.$$

Then $\lim_{n\to\infty} f_n(x)$ is defined for all $x \in \mathbb{R}$ and the *limit function* f is

$$f: \begin{cases} x \mapsto 0 & \text{if } |x| < 1, \\ \pm 1 \mapsto \tfrac{1}{2}, & \\ x \mapsto 1 & \text{if } |x| > 1. \end{cases}$$

The sequence (f_n) does *not* converge uniformly to f on $\mathbb{R}$. (If it did then, using a theorem which states that if (f_n) is a sequence of continuous functions on T which tend to f uniformly on T then f is continuous on T, we could deduce that the limit function is continuous which it manifestly is not.)

The sequence (f_n) is *uniformly bounded* on $\mathbb{R}$, since $\forall x \in \mathbb{R}$, $n \in \mathbb{N}$ we have $|f_n(x)| \leqslant 1$. Thus, 1 is a *uniform bound* for (f_n).

Limit superior and limit inferior

Given a sequence (a_n), $a_n \in \mathbb{R}$, and a number $U \in \mathbb{R}$ such that

(i) $\forall \epsilon > 0$, $\exists N \in \mathbb{N}$ such that $n > N$ implies $a_n < U + \epsilon$,

and (ii) $\forall \epsilon > 0$, $m > 0$, $\exists n > m$ such that $a_n > U - \epsilon$,

then U is called the **limit superior** of (a_n) and we write

$$U = \limsup_{n\to\infty} a_n \; (= \lim_{n\to\infty} \text{lub } a_n).$$

Limit inferior of (a_n) is defined to be $-\limsup_{n\to\infty}(-a_n)$.

See also: Bolzano's Theorem (p. 198); Bolzano–Weierstrass Theorem (p. 198); l'Hospital's Rule (p. 204).

25 Differentiable functions of one variable

We say that the function $f: I \to \mathbb{R}$, where I is an *open* interval, has the **differential coefficient (derivative)** l at $a \in I$ if

$$\lim_{x \to a} \frac{f(x)-f(a)}{x-a} = l.$$

f is then said to be **differentiable** at a.

Note. (i) An equivalent condition is that

$$\lim_{h \to 0} \frac{f(a+h)-f(a)}{h} = l,$$

(ii) by considering $\lim\limits_{x \to a+}$ and $\lim\limits_{x \to a-}$ it is possible to define **right** and **left differential coefficients** at a.

If f is differentiable at each $a \in I$, then f is said to be **differentiable** (or **derivable** (see note on p. 129)) on I.

If f is differentiable on I, we can form a new function $I \to \mathbb{R}$ which maps $a \in I$ onto the differential coefficient of f at a. This new function is called the **derived function** or **derivative** of f and is denoted by f', $\dot{f}$, $\mathrm{d}f/\mathrm{d}x$ (see note on p. 125) or $\mathrm{D}f$. Thus

$$f'(a) = \lim_{x \to a} \frac{f(x)-f(a)}{x-a}.$$

Note. The set of all differentiable functions on I, $\mathscr{D}(I)$, forms a subalgebra of $\mathscr{C}(I)$, that is, sums, products and scalar multiples of differentiable functions are differentiable. Moreover, the function $\mathrm{D}: \mathscr{D}(I) \to \mathbb{R}^I$, defined by $f \mapsto \mathrm{D}f$, is a linear transformation of vector spaces (p. 39).

Example. (i) Let $f: \mathbb{R} \to \mathbb{R}$ be defined by

$$f(x) = \begin{cases} x^2 \sin 1/x, & x \neq 0, \\ 0 & x = 0. \end{cases}$$

Then $$\lim_{x \to 0} \frac{f(x)-f(0)}{x-0} = \lim_{x \to 0} \frac{x^2 \sin 1/x - 0}{x} = \lim_{x \to 0} x \sin 1/x = 0$$

(cf. example (ii), p. 117). Hence, f is *differentiable* at 0 and has *differential coefficient* 0 there.

The *derived function* of f is f' (or $\mathrm{D}f$) : $\mathbb{R} \to \mathbb{R}$ defined by

$$f'(x) = \begin{cases} 2x \sin 1/x - \cos 1/x, & x \neq 0, \\ 0 & x = 0. \end{cases}$$

We note that although f' is defined everywhere on $\mathbb{R}$ it is not continuous at $x = 0$.

(ii) $g: \mathbb{R} \to \mathbb{R}$ defined by $g(x) = |x|$, is not *differentiable* at 0. It has *right* and *left derivatives* there (1 and -1 respectively). It is, however, *differentiable* on any interval excluding 0. If I is a subset of $\{x \in \mathbb{R} | x > 0\}$ then the *derived function* of g on I is $Dg: x \mapsto 1$, if I is a subset of $\{x \in \mathbb{R} | x < 0\}$ then Dg is the function $x \mapsto -1$.

The definitions given above do not generalise readily to functions of several variables having more general domains and codomains. For that reason the following alternative definitions are now preferred.

We say that $f: I \to \mathbb{R}$ is **differentiable** at $a \in I$ provided there exists a *linear mapping* $t: \mathbb{R} \to \mathbb{R}$ such that

$$\lim_{h \to 0} \frac{f(a+h) - f(a) - t(h)}{h} = 0.$$

Alternatively: '...such that

$$f(a+h) - f(a) = t(h) + o(h),$$

where $o(h) \in \mathbb{R}$ and $o(h)/h \to 0$ as $h \to 0$', or '...such that the functions $f(x)$ and $f(a) + t(x-a)$ are "tangent" at a'.

Note. The term *linear mapping* is commonly used to denote two different types of function, namely, (i) (as here) a linear transformation of the kind described on p. 39 (which will always map the zero element onto zero) and (ii) a function which is represented graphically by a straight line (a function $\mathbb{R} \to \mathbb{R}$ of the form $x \mapsto ax + b$). To avoid ambiguity the latter type of function is sometimes referred to as an **affine function**.

The linear mapping t which, when it exists, is unique is then called the **differential** (or **derivative**) of f at a and is denoted by $d_a f$.

We note that in this second definition the differential (derivative) of f at a is not a *number*, an element of $\mathbb{R}$, but a *linear mapping*, i.e. an element of $\mathrm{Hom}(\mathbb{R}, \mathbb{R})$ (p. 41). This apparent anomaly is resolved by our remembering that $\mathrm{Hom}(\mathbb{R}, \mathbb{R})$ is isomorphic to the set of matrices $M_1(\mathbb{R})$ (p. 42), and, hence, to $\mathbb{R}$ itself (for any linear mapping $\mathbb{R} \to \mathbb{R}$ has the form $x \mapsto \alpha x$ for some $\alpha \in \mathbb{R}$). Thus when considering functions from I to $\mathbb{R}$, the differential at a point can be *represented* by a number (i.e. we represent $t: x \mapsto \alpha x$ by α $(= t(1))$, which is, of course, the number yielded by the first definition). This 'coefficient' of the differential is naturally termed the *differential coefficient* at a.

The advantage of this definition is that it generalises readily to functions mapping an open subset A of a Banach space E (p. 111) into a Banach space F. In particular, it carries over to the case where f is a function from a subset of $\mathbb{R}^m$ into $\mathbb{R}^n$. In this case the **differential** or **total derivative** of f at a point $a \in \mathbb{R}^m$ will be a linear mapping $t: \mathbb{R}^m \to \mathbb{R}^n$ which can be uniquely represented (relative to preassigned bases (p. 41)) by an $n \times m$ matrix having coefficients in $\mathbb{R}$. The elements of the matrix will be the *partial differential coefficients* of f at a relative to the chosen bases (see § 26).

Examples. (i) Let $f: \mathbb{R} \to \mathbb{R}$ be defined by $f(x) = 3x^2 + x$. Then at $2 \in \mathbb{R}$ one has

$$f(2+h) - f(2) = 3(4+4h+h^2)+2+h-3.4-2$$
$$= 13h + 3h^2.$$

We note that
$$\lim_{h \to 0} \frac{f(2+h)-f(2)-13h}{h} = 0,$$

and so the required *linear mapping* is $t: h \to 13h$, the *differential* $d_2 f$ is $x \mapsto 13x$, and the *differential coefficient* of f at 2 is 13.

(ii) Let g be defined by $g(x) = 3\sin x$. Then at 0 we have

$$g(0+h) - g(0) = 3\sin h - 0 = 3\sin h.$$

Since
$$\lim_{h \to 0} \frac{3\sin h - 3h}{h} = 0,$$

it follows that the required *linear mapping* is $t: h \mapsto 3h$, the *differential* $d_0 g$ is $x \mapsto 3x$, and the *differential coefficient* of g at 0 is 3.

When the function $f: I \to \mathbb{R}$ is differentiable at every point of I, we say that f is **differentiable** on I. The function $x \mapsto d_x f(1)$ is called the **derivative** of f in I and is denoted by f' or Df.

Note. (i) This latter definition ensures that our two definitions of f' agree. However, an alternative definition would be to define the *derivative* of f to be $f': x \mapsto d_x f$. We should then have $\mathrm{Hom}(\mathbb{R}, \mathbb{R})$ as the codomain of f'. This alternative again has the advantage of being simpler to generalise.

(ii) It is customary in traditional texts to introduce the *differentials* df and dx and to obtain relations such as

$$df = \frac{df}{dx} dx.$$

Using the modern notation this relation would be written

$$d_a f = f'(a) d_a \mathscr{I}$$

where $\mathscr{I}$ (= *id*) denotes the identity function $x \mapsto x$.

Let $f: I \to \mathbb{R}$ be differentiable on I with derivative $f': I \to \mathbb{R}$. f' will or will not belong to $\mathscr{D}(I)$. If it does, then we denote its derivative by f'', $\ddot{f}$, $D^2 f$ or $d^2 f/dx^2$ and refer to it as the **second derivative** of f. In a similar way, we can say that f possesses a **third** or **higher derivatives**. The nth derivative of f is defined to be $D(D^{n-1}f)$, whenever this exists, and is denoted by $f^{(n)}$, $D^n f$ or $d^n f/dx^n$.

We denote the set of all functions $f \in \mathbb{R}^I$ for which the first n derivatives exist and are continuous by $\mathscr{C}^n(I)$ – the set of **n-times continuously differentiable functions,** i.e.

$$\mathscr{C}^1(I) = \{f \in \mathscr{D}(I) | \mathrm{D}f \in \mathscr{C}(I)\} \quad \text{(p. 122)},$$

$$\mathscr{C}^{n+1}(I) = \{f | \mathrm{D}f \in \mathscr{C}^n(I)\}.$$

We then have $\mathscr{C}^n(I) \subset \mathscr{C}^{n-1}(I)$ ($n = 1, 2, \ldots$) and $\mathscr{C}^n(I)$ is a subalgebra of $\mathscr{C}(I)$ ($= \mathscr{C}^0(I)$) (p. 39). We define $\mathscr{C}^\infty(I)$ by

$$\mathscr{C}^\infty(I) = \bigcap_{n=1}^{\infty} \mathscr{C}^n(I).$$

A function belonging to $\mathscr{C}^1(I)$ is said to be **continuously differentiable.**

Examples. (i) We note that the function $f: \mathbb{R} \to \mathbb{R}$ defined in example (i) on p. 122 possesses a derivative f' but that $f' \notin \mathscr{D}(\mathbb{R})$ since $f' \notin \mathscr{C}(\mathbb{R})$. It follows that f does not possess a *second derivative* on $\mathbb{R}$ (although it does so on certain intervals of $\mathbb{R}$). Thus $f \in \mathscr{C}(\mathbb{R})$ and $f \in \mathscr{D}(\mathbb{R})$, but $f \notin \mathscr{C}^1(\mathbb{R})$.

(ii) Let g be defined by $g(x) = 3\sin x$ (cf. example (ii), p. 124). Then

$$\mathrm{D}^n g(x) = \frac{\mathrm{d}^n g}{\mathrm{d}x^n} = 3\sin(x + \tfrac{1}{2}n\pi) \quad \text{and} \quad g \in \mathscr{C}^\infty(\mathbb{R}).$$

Note. The symbols $\frac{\mathrm{d}g}{\mathrm{d}x}, \frac{\mathrm{d}^2 g}{\mathrm{d}x^2}$, etc. and their analogues in the theory of differentiation of functions of several variables are traditionally used in two senses: to denote (i) a function (p. 122) and (ii) its value corresponding to the argument x (example (ii) above). This can lead to confusion and the use of the D notation is accordingly to be preferred. Nevertheless, the use of the traditional notation is still so widespread that we have chosen to frame many definitions in §§26 and 35 in terms of it.

See also: Implicit Function Theorem (p. 205); Inverse Function Theorem (p. 205); Leibniz Formula (p. 207); Lipschitz condition (p. 208); Mean Value Theorems (p. 208); Rolle's Theorem (p. 209); Rolle's conditions (p. 209); Taylor's Theorem (p. 211).

26 Functions of several real variables

Let **f** be a function from A, an open subset of $\mathbb{R}^m$, to $\mathbb{R}^n$. Then, using the notation of p. 16, **f** can be described by the n-tuple $(f_1, \ldots, f_n)$ where f_i denotes the function $\mathrm{pr}_i \circ \mathbf{f} : A \to \mathbb{R}$ $(i = 1, \ldots, n)$. (For this reason, functions such as **f** are often referred to as **vector-valued functions** and are denoted by symbols printed in bold type so as to distinguish them from real-valued or **scalar** functions.)

Since **f** can be expressed in terms of real-valued functions we first consider these. Let $g : A \to \mathbb{R}$, $\mathbf{x} = (x_1, \ldots, x_m) \mapsto g(\mathbf{x})$, be such a function.

Note. In order to help distinguish between points of $\mathbb{R}^m$ $(m \geq 2)$ and their coordinates we shall denote the former by symbols printed in bold type.

We define the ith **partial derivative** of g (or the ith **partial differential coefficient** of g (see note on p. 123) or the **partial derivative of** g **with respect to the** i**th coordinate**) at $\mathbf{a} = (a_1, \ldots, a_m)$ to be

$$\lim_{h \to 0} \frac{g(a_1, \ldots, a_{i-1}, a_i + h, a_{i+1}, \ldots, a_m) - g(a_1, \ldots, a_m)}{h},$$

whenever that limit exists, and denote it by

$$\mathrm{D}_i g|_{\mathbf{a}}, \quad \mathrm{D}_i g(\mathbf{a}), \quad g_{x_i}(\mathbf{a}) \quad \text{or} \quad \left.\frac{\partial g}{\partial x_i}\right|_{\mathbf{a}}.$$

If the ith partial derivative is defined at all $\mathbf{a} \in A$, then we refer to the function $A \to \mathbb{R}$, $\mathbf{a} \mapsto \mathrm{D}_i g|_{\mathbf{a}}$ as the ith **partial derivative of** g and denote it by

$$\mathrm{D}_i g, \quad g_{x_i} \quad \text{or} \quad \frac{\partial g}{\partial x_i}.$$

Such derivatives are termed **first order partial derivatives.** Partial derivatives of higher orders are obtained, when they exist, by further differentiation. Typical **higher order partial derivatives** are (in the various notations):

$$\frac{\partial^2 g}{\partial x_j \, \partial x_i} \left(= \frac{\partial}{\partial x_j}\left(\frac{\partial g}{\partial x_i}\right)\right), \qquad g_{x_i x_i} = \mathrm{D}_i^2 g \ (= \mathrm{D}_i(\mathrm{D}_i g)).$$

Note. It can be proved that if $D_{i,j}g\ (= D_i(D_jg))$ and $D_{j,i}g$ exist and are continuous then they are equal.

Example. Let $g: \mathbb{R}^2 \to \mathbb{R}$ be defined by $(x, y) \mapsto x^2y + 3y^2x$.

Then

$$\left.\frac{\partial g}{\partial x}\right|_{(a,b)} = \lim_{h\to 0} \frac{(a+h)^2b + 3b^2(a+h) - a^2b + 3b^2a}{h} = 2ab + 3b^2.$$

Thus the *partial differential coefficient* of g with respect to x at (a, b) is $2ab+3b^2$. The *partial differential coefficient* of g with respect to y at (a, b) is a^2+6ab.

We have (see the note on p. 125)

$$\frac{\partial g}{\partial x} = 2xy + 3y^2, \quad \frac{\partial g}{\partial y} = x^2 + 6xy,$$

$$\frac{\partial^2 g}{\partial x^2} = 2y, \quad \frac{\partial^2 g}{\partial y\,\partial x} = 2x + 6y = \frac{\partial^2 g}{\partial x\,\partial y}, \quad \frac{\partial^2 g}{\partial y^2} = 6x.$$

Given a function $g: A \to \mathbb{R}$ possessing first order partial derivatives we can associate with $\mathbf{a} \in A$ the vector

$$\left(\left.\frac{\partial g}{\partial x_1}\right|_{\mathbf{a}}, \left.\frac{\partial g}{\partial x_2}\right|_{\mathbf{a}}, \ldots, \left.\frac{\partial g}{\partial x_m}\right|_{\mathbf{a}}\right)$$

whose components are the partial derivatives of g at $\mathbf{a}$.

Similarly, a function $\mathbf{f}: A \to \mathbb{R}^n$, whose component functions possess first order partial derivatives, can be associated with the $n \times m$ matrix

$$\left(\left.\frac{\partial f_i}{\partial x_j}\right|_{\mathbf{a}}\right) \quad (i = 1, \ldots, n;\ j = 1, \ldots, m).$$

This matrix is known as the **Jacobian matrix** of $\mathbf{f}$ at $\mathbf{a}$, and is denoted by $J_{\mathbf{f}}(\mathbf{a})$. When $m = n$, the determinant of the Jacobian matrix of $\mathbf{f}$ is called the **Jacobian** of $\mathbf{f}$. The Jacobian is traditionally denoted by

$$\frac{\partial(f_1, f_2, \ldots, f_m)}{\partial(x_1, x_2, \ldots, x_m)} \quad \text{(see p. 133)}.$$

Warning. This latter symbol $\mathcal{J}_f(\mathbf{a})$ are also used to denote the Jacobian matrix and the Jacobian respectively. Additional confusion ensues when the same signs are used to denote the determinant of a matrix and the absolute value of a number.

Example. Let $\mathbf{f}: A \times B \to \mathbb{R}^2$ be defined as in the example on p. 16. Then we have $f_1: (r, \theta) \mapsto r\cos\theta$, $f_2: (r, \theta) \mapsto r\sin\theta$.

Corresponding to the function f_1 we have the vector

$$\left(\frac{\partial f_1}{\partial r}\bigg|_{(r,\theta)}, \frac{\partial f_1}{\partial \theta}\bigg|_{(r,\theta)}\right) = (\cos\theta, -r\sin\theta)$$

and corresponding to f_2, we have the vector

$$\left(\frac{\partial f_2}{\partial r}\bigg|_{(r,\theta)}, \frac{\partial f_2}{\partial \theta}\bigg|_{(r,\theta)}\right) = (\sin\theta, r\cos\theta).$$

The *Jacobian matrix* of $\mathbf{f}$ at (r, θ) is

$$J_{\mathbf{f}}(r,\theta) = \begin{pmatrix} \cos\theta & -r\sin\theta \\ \sin\theta & r\cos\theta \end{pmatrix}$$

and the *Jacobian* of $\mathbf{f}$ at (r, θ) is

$$|\mathcal{J}_{\mathbf{f}}(r,\theta)| = \frac{\partial(f_1, f_2)}{\partial(r, \theta)} = r.$$

There are certain important differences between functions of one variable and functions of several variables. In the case of functions of one variable, the existence of a derivative at a point implies that the function is continuous at that point. The possession of partial derivatives at a point does not, however, imply the continuity of a function of several variables at that point – more is required.

Again, in the case of functions of one variable, the existence of a derivative at a implies that it is possible to approximate to the function f in a neighbourhood of a by means of a linear function (indeed this notion formed the basis for the definition of the differential of f). For the first definition of a derivative (p. 122) leads naturally to the approximation

$$\delta y = f(a+\delta x) - f(a) \simeq f'(a)\,\delta x$$

(where δx denotes some small number – an *increment* in the argument of f (p. 13) – and δy the corresponding increment in the value of f), or, equivalently, to the equation

$$f(a+h) - f(a) = f'(a)h + o(h)$$

on which the second definition was based.

A function of one variable possessing a derivative at a point is, therefore, in the above sense, *locally linear* there. The concept of local linearity can be extended to functions of several variables by saying that $g: A \to \mathbb{R}$ is **locally linear** at $\mathbf{a} \in A$ if $\alpha_1, \ldots, \alpha_m \in \mathbb{R}$ can be found such that

$$g(a_1+h_1, \ldots, a_m+h_m) - g(a_1, \ldots, a_m) \simeq \alpha_1 h_1 + \alpha_2 h_2 + \ldots + \alpha_m h_m$$

for all sufficiently small $h_1, \ldots, h_m$ (or, more precisely, if a linear transformation $t: \mathbb{R}^m \to \mathbb{R}$ can be found such that

$$g(\mathbf{a}+\mathbf{x})-g(\mathbf{a}) = t(\mathbf{x})+o(\mathbf{x})$$

for all $\mathbf{x}$ in some open neighbourhood of the origin such that $\mathbf{a}+\mathbf{x} \in A$, where $o(\mathbf{x})$ is small in the sense that

$$\lim_{\mathbf{x}\to\mathbf{0}} \frac{\|o(\mathbf{x})\|}{\|\mathbf{x}\|} = 0 \quad \text{(p. 111)).}$$

It is, however, no longer true that possession of first order partial derivatives ensures local linearity (although the latter property does imply the former). It is necessary, therefore, to distinguish between these two properties in the case of functions of two or more variables.

If $g: A \to \mathbb{R}$ is locally linear at $\mathbf{a} \in A$, then we say that g is **differentiable** at $\mathbf{a}$ and (in an analogous manner to the case $g: I \to \mathbb{R}$) we say that the linear transformation $t: \mathbb{R}^m \to \mathbb{R}$ is the **differential** of g at $\mathbf{a} \in A$, and we denote it by $\mathrm{d}_\mathbf{a} g$.

If $g: A \to \mathbb{R}$ is differentiable at every point $\mathbf{a} \in A$, then we say it is **differentiable** on A.

(It is easily shown that if g is differentiable at $\mathbf{a}$, then the numbers $\alpha_1, \ldots, \alpha_m$ referred to above are, in fact, $\mathrm{D}_1 g|_\mathbf{a}, \ldots, \mathrm{D}_m g|_\mathbf{a}$.)

Note. (i) A function g which is differentiable at a point $\mathbf{a}$ will be continuous at $\mathbf{a}$. A function g is differentiable at $\mathbf{a}$ provided the m first order partial derivatives are continuous at $\mathbf{a}$.

(ii) It is in order to bring out this difference between 'possessing derivatives' and being 'differentiable' that some authors prefer to describe functions of one variable as *derivable* (p. 122) if it is shown *using the traditional definition* that they possess a derivative everywhere in their domain, even though, in the case of such functions, *derivable* and *differentiable* are equivalent.

$\mathrm{d}_\mathbf{a} g \in \mathrm{Hom}(\mathbb{R}^m, \mathbb{R})$, an m-dimensional vector space over $\mathbb{R}$ (p. 41), and the coefficients of $\mathrm{d}_\mathbf{a} g$ with respect to the usual basis for $\mathbb{R}^m$ are the m partial differential coefficients

$$\left.\frac{\partial g}{\partial x_1}\right|_\mathbf{a}, \quad \ldots, \quad \left.\frac{\partial g}{\partial x_m}\right|_\mathbf{a},$$

or, in what are now becoming the preferred notations for advanced work,

$$\mathrm{D}_1 g|_\mathbf{a}, \ \ldots, \ \mathrm{D}_m g|_\mathbf{a} \ \text{ or } \ \mathrm{D}_1 g(\mathbf{a}), \ \ldots, \ \mathrm{D}_m g(\mathbf{a}).$$

Denoting the elements of the basis for $\mathbb{R}^m$ by $\mathbf{e}_1, \ldots, \mathbf{e}_m$, we have

$$\mathrm{D}_i g|_\mathbf{a} = \mathrm{d}_\mathbf{a} g(\mathbf{e}_i) \quad (i = 1, \ldots, m).$$

Examples. (i) Let

$$f(x_1, x_2) = \begin{cases} \dfrac{x_1 x_2}{x_1^2 + x_2^2} & (x_1, x_2) \neq (0, 0), \\ 0 & (x_1, x_2) = (0, 0). \end{cases}$$

Then $D_1 f|_{(0,0)} = D_2 f|_{(0,0)} = 0$, i.e. the function has partial derivatives with respect to x_1 and x_2 at the origin.

f is *not* differentiable at the origin, however, for we cannot find any linear mapping $t : \mathbb{R}^2 \to \mathbb{R}$ (of the form $(x_1, x_2) \mapsto \alpha x_1 + \beta x_2$) such that

$$f(x_1, x_2) - 0 = t(x_1, x_2) + o(x_1, x_2).$$

To see this we consider what happens when we approach $(0, 0)$ along the line $x_1 = x_2$. We then have

$$f(x_1, x_2) = \frac{x_1^2}{x_1^2 + x_1^2} = \frac{1}{2} \quad \text{and as} \quad (x_1, x_2) \to (0, 0), \quad f(x_1, x_2) \to \tfrac{1}{2}.$$

Since, if suitable t and $o(\mathbf{x})$ existed, $t(x_1, x_2) + o(x_1, x_2)$ must tend to 0 as $(x_1, x_2) \to \mathbf{0}$ it follows that no such t can be found.

(ii) Let $g : \mathbb{R}^2 \to \mathbb{R}$ be defined by $(x, y) \mapsto x^2 y + 3y^2 x$ (cf. example p. 127). Then taking $\mathbf{a} = (1, 1)$ we have

$$\begin{aligned} g(\mathbf{a} + \mathbf{x}) - g(\mathbf{a}) &= (x^2 + 2x + 1)(y + 1) + 3(y^2 + 2y + 1)(x + 1) - 4 \\ &= 5x + 7y + x^2 y + 8xy + 3y^2 x + x^2 + 3y^2. \end{aligned}$$

Hence, putting

$$t(\mathbf{x}) = 5x + 7y \quad \text{and} \quad o(\mathbf{x}) = x^2 y + 8xy + 3y^2 x + x^2 + 3y^2,$$

we see, since
$$\frac{\|o(\mathbf{x})\|}{\|\mathbf{x}\|} \to 0 \quad \text{as} \quad \mathbf{x} \to \mathbf{0},$$

that g is *differentiable* at $(1, 1)$. $d_{\mathbf{a}} g$, the *differential* of g at $\mathbf{a} = (1, 1)$, is the map $(x, y) \mapsto 5x + 7y$. Note that

$$d_{\mathbf{a}} g(\mathbf{e}_1) = 5 = D_1 g|_{\mathbf{a}} \quad \text{and} \quad d_{\mathbf{a}} g(\mathbf{e}_2) = 7 = D_2 g|_{\mathbf{a}}.$$

Note. A traditional definition of the **(total) differential** of g would be (compare p. 124)

$$dg = \frac{\partial g}{\partial x_1} dx_1 + \frac{\partial g}{\partial x_2} dx_2 + \ldots + \frac{\partial g}{\partial x_m} dx_m,$$

an expression whose equivalent in the alternative notation is

$$d_{\mathbf{a}} g = D_1 g|_{\mathbf{a}} d_{\mathbf{a}} \pi^1 + D_2 g|_{\mathbf{a}} d_{\mathbf{a}} \pi^2 + \ldots + D_m g|_{\mathbf{a}} d_{\mathbf{a}} \pi^m$$

where π^i ($= \mathrm{pr}_i$ (p. 11)) denotes the function $\mathbf{x} \mapsto x_i$.

Returning to the consideration of the more general function $\mathbf{f} : A \to \mathbb{R}^n$ (where A is again an open subset of $\mathbb{R}^m$) we find that an analogous definition to that given above leads to the definition of $d_{\mathbf{a}} \mathbf{f}$, the **differential** of $\mathbf{f}$ at $\mathbf{a}$, as a linear transformation from $\mathbb{R}^m$ to $\mathbb{R}^n$. Relative to the usual bases, $d_{\mathbf{a}}\mathbf{f}$ is represented by the matrix

$$(D_j(\pi^i \circ \mathbf{f})|_{\mathbf{a}}) \quad (i = 1, \ldots, n; j = 1, \ldots, m)$$

$$= (D_j f_i(\mathbf{a})) = \left(\frac{\partial f_i}{\partial x_j}\bigg|_{\mathbf{a}}\right)$$

i.e. the Jacobian matrix (p. 127) of $\mathbf{f}$ at $\mathbf{a}$.

Example. Let $f: A \times B \to \mathbb{R}^2$ be defined as in the example on p. 127. At the point $\mathbf{a} = (r, \theta)$ we have

$$d_{\mathbf{a}} f_1 : (u, v) \mapsto (u \cos\theta, \ -rv \sin\theta),$$

$$d_{\mathbf{a}} f_2 : (u, v) \mapsto (u \sin\theta, rv \cos\theta).$$

and $d_{\mathbf{a}}\mathbf{f}$, the *differential* of $\mathbf{f}$ at $\mathbf{a}$, is the map $A \times B \to \mathbb{R}^2$ represented (relative to the usual bases) by the matrix

$$\begin{pmatrix} \cos\theta & -r\sin\theta \\ \sin\theta & r\cos\theta \end{pmatrix}.$$

In this particular example

$$D_1(\pi^1 \circ \mathbf{f})|_{(r,\theta)} = \cos\theta, \quad D_2(\pi^1 \circ \mathbf{f})|_{(r,\theta)} = -r\sin\theta, \quad \text{etc.}$$

As was mentioned above, knowing that a function possesses partial derivatives is less informative than knowing it is differentiable, for the former only provides us with knowledge of the function's behaviour in *certain* 'directions'. If a function is differentiable, then it will have a derivative in *any* direction. We say that the **derivative of $g: A \to \mathbb{R}$ at a in the direction $\mathbf{v} \in \mathbb{R}^m$** is

$$\lim_{h \to 0} \left(\frac{g(\mathbf{a}+h\mathbf{v}) - g(\mathbf{a})}{h\|\mathbf{v}\|}\right).$$

Thus in the special case when $m = 1$ we obtain the derivative of g, while, in general, if $\mathbf{v} = \mathbf{e}_i$ (for some $i = 1, \ldots, m$) then the derivative in the direction $\mathbf{v}$ is the ith partial derivative.

Coordinate systems

It is frequently necessary to refer to 'coordinate systems' other than the 'usual' system in $\mathbb{R}^m$ and to obtain partial differential coefficients of a function $g: A \to \mathbb{R}$ in terms of the 'new coordinates'. A **coordinate system** on a subset $A \subset \mathbb{R}^m$ is defined to be a bijection $\mu: A \to B$, where B is also a subset of $\mathbb{R}^m$, such that both μ and μ^{-1}

are differentiable at every point of A and B respectively. The partial derivatives of $\mathbf{f} \circ \boldsymbol{\mu}$ can then be obtained from those of $\mathbf{f}$ by means of the *chain rule* (p. 200).

Coordinate systems of particular interest include:

(1) the non-singular linear transformations (*affine transformations*)

$$T: \mathbb{R}^m \to \mathbb{R}^m;$$

(2) **polar coordinates** in $\mathbb{R}^2$ (p. 69) given by

$$\boldsymbol{\mu}: (x, y) \mapsto (r, \theta)$$

where
$$r = \sqrt{(x^2+y^2)} \quad (r > 0),$$
$$\theta = \arg(x+iy) \quad (0 \leqslant \theta < 2\pi)$$

(note that $\boldsymbol{\mu}$ is not defined everywhere in $\mathbb{R}^2$ and that $\boldsymbol{\mu}^{-1}$ is not continuous everywhere in $\mathbb{R}^2$);

(3) **polar coordinates** in $\mathbb{R}^3$:

(*a*) **spherical polar coordinates** given by

$$(x, y, z) \mapsto (r, \theta, \phi)$$

where
$$r = \sqrt{(x^2+y^2+z^2)} \quad (r > 0),$$
$$\theta = \cos^{-1}\frac{x}{\sqrt{(x^2+y^2)}} \quad (0 \leqslant \theta < 2\pi),$$
$$\phi = \cos^{-1}\frac{z}{\sqrt{(x^2+y^2+z^2)}} \quad (0 < \phi < \pi);$$

(*b*) **cylindrical polar coordinates** given by

$$(x, y, z) \mapsto (\rho, \theta, z)$$

where
$$\rho = \sqrt{(x^2+y^2)} \quad (\rho > 0),$$
$$\theta = \cos^{-1}\frac{x}{\sqrt{(x^2+y^2)}} \quad (0 \leqslant \theta < 2\pi),$$
$$z = z.$$

(Again, the domain and codomain of these coordinate systems are proper subsets of $\mathbb{R}^3$.)

Example. Let $\mathbf{f}: A\,(\subset \mathbb{R}^3) \to B\,(\subset \mathbb{R}^3)$ be defined by

$$\mathbf{f}: (x, y, z) \mapsto (x^2+y^2+z^2, \log(x^2+y^2+z^2), e^{x^2+y^2+z^2}).$$

Then 'converting' to *spherical polar coordinates* in $\mathbb{R}^3$, i.e. defining $\boldsymbol{\mu}: (x, y, z) \mapsto (r, \theta, \phi)$ as in 3(*a*) above, and setting

$$\mathbf{g}: (r, \theta, \phi) \mapsto (r^2, \log r^2, e^{r^2}),$$

we have $\mathbf{f} = \mathbf{g} \circ \boldsymbol{\mu}$. It follows by the chain rule (p. 200) that

$$J_{\mathbf{f}}(\mathbf{x}) = J_{\mathbf{g}}(\boldsymbol{\mu}(\mathbf{x}))\,J_{\boldsymbol{\mu}}(\mathbf{x})$$

(or, in traditional notation,

$$\frac{\partial(f_1, f_2, f_3)}{\partial(x, y, z)} = \frac{\partial(f_1, f_2, f_3)}{\partial(r, \theta, \phi)}\,\frac{\partial(r, \theta, \phi)}{\partial(x, y, z)}$$

(a formula which governed the choice of notation)).

Hence, setting $\mathbf{x} = (x, y, z)$ and $r = \sqrt{(x^2+y^2+z^2)}$, we have

$$\begin{pmatrix} D_1f_1(\mathbf{x}) & D_2f_1(\mathbf{x}) & D_3f_1(\mathbf{x}) \\ D_1f_2(\mathbf{x}) & D_2f_2(\mathbf{x}) & D_3f_2(\mathbf{x}) \\ D_1f_3(\mathbf{x}) & D_2f_3(\mathbf{x}) & D_3f_3(\mathbf{x}) \end{pmatrix} = \begin{pmatrix} 2r & 0 & 0 \\ \dfrac{2}{r} & 0 & 0 \\ 2re^{r^2} & 0 & 0 \end{pmatrix} \begin{pmatrix} \dfrac{x}{r} & \dfrac{y}{r} & \dfrac{z}{r} \\ * & * & * \\ * & * & * \end{pmatrix}$$

$$= \begin{pmatrix} 2x & 2y & 2z \\ \dfrac{2x}{r^2} & \dfrac{2y}{r^2} & \dfrac{2z}{r^2} \\ 2xe^{r^2} & 2ye^{r^2} & 2ze^{r^2} \end{pmatrix}.$$

(The terms indicated by an asterisk have not been inserted, since in this example the elements of the product matrix do not depend upon them.)

See also: Chain rule (p. 200); Heine's Theorem (p. 204); Implicit Function Theorem (p. 205); Inverse Function Theorem (p. 205); Mean Value Theorems (p. 208); Taylor series (p. 212).

27 Integration

The integral of a function can be defined in several ways; each deeper definition encompassing a yet more varied class of function. We give below three definitions – those of the *Cauchy*, *Riemann*, and *Riemann–Stieltjes* integrals – and include a fourth definition – that of the *Lebesgue* integral – in a later section.

The Cauchy integral

Let f be a function from a closed finite interval $I = \langle a, b \rangle$ of $\mathbb{R}$ $(a \neq b)$ into $\mathbb{R}$. We say that a continuous mapping g of I into $\mathbb{R}$ is a **primitive** of f in I if there exists a countable set $D \subset I$ such that, for any $x \in I - D$, g is differentiable at x and $g'(x) = f(x)$. It follows that if g_1 and g_2 are any two primitives of f, then $g_1 - g_2$ is a constant function (p. 13), on I. The set of all primitives of f is called the **indefinite integral** of f, and the particular primitive g for which $g(a) = 0$ is called the **definite integral** of f and is denoted by

$$g(x) = \int_a^x f(t)\,dt \quad \left(\text{or, alternatively, by} \int_a^x f\right).$$

The number $g(b) - g(a)$ is independent of the choice of primitive g and is called the **integral of** f **from** a **to** b (or, **between** a and b). It is denoted, as occasion demands, by

$$\int_a^b f(t)\,dt, \quad \int_a^b f \quad \text{or} \quad [g(x)]_a^b.$$

Note. (i) The choice of t in the symbol for the integral is not significant. It is a '*dummy variable*' and could equally well be replaced by, say, y or ξ.

(ii) f can be shown to possess a primitive, and hence a Cauchy integral, if it is a regulated function (p. 118). In particular, if f is continuous on I, then there exists a primitive g which is differentiable on $\langle a, b \rangle$ and which satisfies $g'(x) = f(x)$ for all $x \in \langle a, b \rangle$ (i.e. the set D in the definition given above can be taken to be the empty set). Step functions (p. 118) will always possess primitives which are known as **piecewise linear functions.**

Examples. (i) Consider the function sine : $\langle 0, b \rangle \to \mathbb{R}$. Then

$$g_1 : x \mapsto \tfrac{1}{2}\pi - \cos x \quad \text{and} \quad g_2 : x \mapsto e - \cos x$$

are two *primitives* of sine in $I = \langle 0, b \rangle$. The set of all functions

$$\{x \mapsto a - \cos x \mid a \in \mathbb{R}\}$$

is the *indefinite integral* of sine. The *definite integral* of sine on I is the primitive defined by

$$g(x) = 1 - \cos x$$

and denoted by

$$\int_0^x \sin t \, dt.$$

The *integral of sine from 0 to b*, written

$$\int_0^b \sin t \, dt \quad \text{or} \quad [1 - \cos x]_0^b$$

is

$$g(b) - g(0) = 1 - \cos b.$$

(ii) Let $f: \leqslant 0, 2 \geqslant \to \mathbb{R}$ be defined by

$$f(x) = \begin{cases} 0 & \text{if } 0 \leqslant x \leqslant 1, \\ 1 & \text{if } 1 < x \leqslant 2. \end{cases}$$

(f is *not* continuous but is *regulated* (it is, in fact, a *step function*).) Then $g: \leqslant 0, 2 \geqslant \to \mathbb{R}$ defined by

$$g(x) = \begin{cases} 0 & \text{if } 0 \leqslant x \leqslant 1, \\ x - 1 & \text{if } 1 < x \leqslant 2, \end{cases}$$

is a continuous function which is differentiable on $< 0, 2 > - \{1\}$ and which satisfies $g'(x) = f(x)$ on $< 0, 2 > - \{1\}$. g is, therefore, a *primitive* of f on $\leqslant 0, 2 \geqslant$ and a *piecewise linear function.*

The Riemann integral

Let f be a bounded function (p. 119) defined on a closed finite interval $I = \leqslant a, b \geqslant$ of $\mathbb{R}$ ($a \neq b$).

Corresponding to each partition $P = \{x_0, \ldots, x_n\}$ of I (p. 118) we put

$$M_i = \operatorname{lub} f(t) \quad (x_{i-1} \leqslant t \leqslant x_i),$$

$$m_i = \operatorname{glb} f(t) \quad (x_{i-1} \leqslant t \leqslant x_i),$$

$$U(P, f) = \sum_{i=1}^{n} M_i \, \Delta x_i \quad \left(= \sum_{i=1}^{n} M_i (x_i - x_{i-1}) \right),$$

$$L(P, f) = \sum_{i=1}^{n} m_i \, \Delta x_i.$$

The glb of $\{U(P, f)\}$ taken over all partitions P of I is then known as the **upper Riemann integral** of f and the corresponding lub of $\{L(P, f)\}$ as the **lower Riemann integral**. If the upper and lower integrals are equal (it can be shown that the former is always greater than or equal to the latter), then we say that f is **Riemann integrable** on I and we denote the common value of the two integrals by

$$\int_a^b f(x)\,dx$$

which we call the **(Riemann) integral of f from a to b.**

Note. (i) It can be shown that if f is bounded on I then it is Riemann integrable (often denoted by $f \in \mathscr{R}$) on I if and only if it is continuous *almost everywhere* (p. 187) on I.

(ii) All functions of bounded variation (p. 118) on I are Riemann integrable on I.

(iii) The **Fundamental Theorem of Calculus** states that if f is Riemann integrable on $I = \leqslant a, b \geqslant$ and g is a differentiable function on $< a, b >$ such that (I) $g'(x) = f(x)$ for all $x \in < a, b >$, and (II) $\lim_{x\to a+} g(x)$ and $\lim_{x\to b-} g(x)$ exist, then

$$\int_a^b f(x)\,dx = \lim_{x\to b-} g(x) - \lim_{x\to a+} g(x).$$

(When using this theorem to evaluate integrals we usually deal with functions g for which $\lim_{x\to a+} g(x) = g(a)$ and $\lim_{x\to b-} g(x) = g(b)$.)

Example. Consider $f: \leqslant a, b \geqslant \to \mathbb{R}$ defined by $f(x) = x$ and let $P = \{x_0, \ldots, x_n\}$ be a partition of $I = \leqslant a, b \geqslant$.

Then

$$M_i = x_i, \quad m_i = x_{i-1} \quad (i = 1, \ldots, n),$$

$$U(P,f) = \sum_{i=1}^{n} x_i(x_i - x_{i-1}),$$

$$L(P,f) = \sum_{i=1}^{n} x_{i-1}(x_i - x_{i-1}).$$

Hence,

$$0 \leqslant U(P,f) - L(P,f) = \sum_{i=1}^{n} (x_i - x_{i-1})(x_i - x_{i-1})$$

$$\leqslant \mu \sum_{i=1}^{n} (x_i - x_{i-1}), \quad \text{where} \quad \mu = \text{lub}\{|x_i - x_{i-1}|\},$$

$$= \mu(b-a).$$

Therefore, $U(P,f) - L(P,f) < \epsilon$ provided $\mu < \epsilon/b-a$. Now

$$2x_{i-1} < x_{i-1} + x_i < 2x_i,$$

and so

$$2x_{i-1}(x_i - x_{i-1}) < x_i^2 - x_{i-1}^2 < 2x_i(x_i - x_{i-1}).$$

Summing we have

$$2L(P,f) < \sum_{i=1}^{n} (x_i^2 - x_{i-1}^2) = b^2 - a^2 < 2U(P,f).$$

Since $U(P,f)-L(P,f) < \epsilon$ provided the mesh (p. 118) of the partition is sufficiently small, it follows that

$$\text{glb}\{U(P,f)\} = \text{lub}\{L(P,f)\} = \frac{b^2-a^2}{2} = \int_a^b x\,dx,$$

i.e. f is *Riemann integrable* on I.

The Riemann–Stieltjes integral

Let f and α be bounded, real-valued functions defined upon a closed finite interval $I = \langle a, b\rangle$ of $\mathbb{R}$ $(a \neq b)$, $P = \{x_0, \ldots, x_n\}$ a partition of I, and t_i a point of the subinterval $\langle x_{i-1}, x_i\rangle$. A sum of the form

$$S(P,f,\alpha) = \sum_{i=1}^{n} f(t_i)\,(\alpha(x_i)-\alpha(x_{i-1}))$$

is called a **Riemann–Stieltjes sum** of f with respect to α. f is said to be **Riemann integrable with respect to** α on I if there exists $A \in \mathbb{R}$ such that given any $\epsilon > 0$ there exists a partition P_ϵ of I for which, for all P finer (p. 118) than P_ϵ and for every choice of points t_i, we have

$$|S(P,f,\alpha)-A| < \epsilon.$$

If such an A exists, then it is unique and is known as the **Riemann–Stieltjes integral of** f **with respect to** α. f is known as the **integrand** and α the **integrator**. The integral is denoted by

$$\int_a^b f\,d\alpha \quad \text{or} \quad \int_a^b f(x)\,d\alpha(x).$$

Note. (i) Alternative definitions of the Riemann–Stieltjes integral can be found in the literature (see (iii) below), not all of which are equivalent.

(ii) Monotonically increasing integrators play a particularly important role in this theory of integration and for that reason many authors restrict the definition of the Riemann–Stieltjes integral to such integrators.

(iii) One alternative definition is based upon the consideration of 'upper' and 'lower' sums as in the definition of the Riemann integral on p. 135. We then have

$$U(P,f,\alpha) = \sum_{i=1}^{n} M_i(\alpha(x_i)-\alpha(x_{i-1})) \quad \text{and so on.}$$

If α is monotonic increasing, then

$$L(P,f,\alpha) \leqslant U(P,f,\alpha).$$

Again, upper and lower integrals are defined in terms of the glb and lub of the upper and lower sums respectively taken over all possible partitions, and f is said to be integrable with respect to α if the two integrals so defined are equal.

(iv) The conditions for $S(P, f, \alpha)$ to possess a limit for 'finer and finer' partitions can be presented in several ways. One common alternative is used in the definition of a contour integral given on p. 156.

(v) When α is the identity function $x \mapsto x$, one obtains the Riemann integral of f as a special case of the Riemann–Stieltjes integral.

(vi) By using as integrator the square bracket function $x \to [x]$ (p. 118), we can write every finite sum as a Riemann–Stieltjes integral. For, given $\sum_{i=1}^{n} a_i$, we define f on $\leqslant 0, n \geqslant$ by

$$f(x) = a_i \quad \text{if} \quad i-1 < x \leqslant i \quad (i = 1, \ldots, n),$$

$$f(0) = 0.$$

We then have
$$\sum_{i=1}^{n} a_i = \int_0^n f(x)\,\mathrm{d}[x].$$

(vii) If f is integrable with respect to α on $I = \leqslant a, b \geqslant$, then α is integrable with respect to f on I and we have

$$\int_a^b f\,\mathrm{d}\alpha + \int_a^b \alpha\,\mathrm{d}f = f(b)\,\alpha(b) - f(a)\,\alpha(a).$$

In particular, when α has a continuous derivative α' on I, then the Riemann integral

$$\int_a^b f(x)\,\alpha'(x)\,\mathrm{d}x$$

exists and is equal to
$$\int_a^b f(x)\,\mathrm{d}\alpha(x).$$

In this case the above relation yields the well-known formula for *integration by parts*.

In the case of all three integrals – Cauchy, Riemann and Riemann–Stieltjes – if $a < b$ and the relevant integrals exist on $\leqslant a, b \geqslant$, we define

$$\text{(i)} \int_b^a \text{ to be } -\int_a^b \quad \text{and} \quad \text{(ii)} \int_a^a = 0.$$

Example. To evaluate
$$\int_0^3 x\,\mathrm{d}(x - [x]).$$

Consider first
$$\int_0^3 (x - [x])\,\mathrm{d}x.$$

The restriction (p. 13) of $f : x \mapsto (x - [x])$ to $\leqslant 0, 1 >$ is the function $f_1 : x \mapsto x$. It follows then from the *Fundamental Theorem of Calculus* (p. 136) that

$$\int_0^1 (x - [x])\,\mathrm{d}x = \int_0^1 x\,\mathrm{d}x = \lim_{x \to 1-} \frac{x^2}{2} - \lim_{x \to 0+} \frac{x^2}{2} = \frac{1}{2}.$$

(Note that this is true even though $\lim_{x \to 1-} f(x) \neq f(1)$.)

If we now consider a typical *Riemann–Stieltjes* sum $S(P, f, \alpha)$ on $\leqslant 1, 2 \geqslant$ (where α is the identity function) we find that it is identical to the sum $S(P', f, \alpha)$ on $\leqslant 0, 1 \geqslant$ when the partition P' is obtained from P by replacing x_i by $x_i - 1$ ($i = 0, \ldots, n$). It follows, therefore, that

$$\int_1^2 (x - [x])\,dx$$

exists and equals $\frac{1}{2}$. Similarly,

$$\int_2^3 (x - [x])\,dx = \tfrac{1}{2}$$

and so
$$\int_0^3 (x - [x])\,dx$$

exists and is equal to $\frac{3}{2}$.

It follows (note (vii) above), that

$$\int_0^3 x\,d(x - [x])$$

exists and that

$$\int_0^3 x\,d(x - [x]) = 3.0 - 0.0 - \int_0^3 (x - [x])\,dx = -\tfrac{3}{2}.$$

See also: Integration by substitution (p. 201); Mean Value Theorems (p. 208); Simpson's rule (p. 210); Wallis's formulae (p. 213).

28 Infinite series and products

Given any sequence of real or complex numbers (a_n) we can form from it a second sequence (A_n), where

$$A_n = \sum_{i=0}^{n} a_i = a_0 + a_1 + \ldots + a_n.$$

If the infinite sequence (A_n) has a limit A as n tends to infinity, then we say that A is the **sum** of the **infinite series**

$$a_0 + a_1 + \ldots + a_n + \ldots = \sum_{n=0}^{\infty} a_n \; (= \Sigma a_n = \sum_{\mathbb{N}} a_n).$$

We say then that the infinite series **converges** to A and that the series is **convergent**. The numbers A_n and a_n are known respectively as the **partial sums** and the **terms** of the infinite series.

If the sequence (A_n) does not have a limit, then we say that the infinite series

$$a_0 + a_1 + \ldots + a_n + \ldots$$

is **divergent**.

An infinite series of real numbers is *properly divergent or oscillatory* (*oscillating*) if the sequence of partial sums is properly divergent or oscillatory (p. 115).

Examples. Particular series of note include:

(i) The **harmonic series**

$$1 + \frac{1}{2} + \frac{1}{3} + \frac{1}{4} + \ldots + \frac{1}{n} + \ldots,$$

which is properly divergent.

(ii) The **alternating harmonic series**

$$1 - \frac{1}{2} + \frac{1}{3} - \frac{1}{4} + \ldots + \frac{(-1)^{n+1}}{n} + \ldots,$$

which is convergent.

(iii) The **geometric series**

$$1 + x + x^2 + x^3 + \ldots + x^n + \ldots,$$

which converges (to $1/1-x$) if and only if $|x| < 1$.

An infinite series of real numbers Σa_n is **alternating** if its terms are alternately positive and negative numbers.

An infinite series of real or complex numbers Σa_n is said to be **absolutely convergent** when $\Sigma |a_n|$ converges.

When Σa_n converges but $\Sigma |a_n|$ does not, then Σa_n is said to be **conditionally convergent.**

Note. If $\Sigma |a_n|$ converges then so does Σa_n, i.e. absolutely convergent series are convergent.

A series is said to be **rearranged** when the order of its terms is changed.

Examples. (i) $\sum_{n=1}^{\infty} \frac{1}{n(n+1)}$ is *convergent* to 1, since the Nth *partial sum*, A_N, is

$$\sum_{n=1}^{N} \frac{1}{n(n+1)} = \sum_{n=1}^{N} \left(\frac{1}{n} - \frac{1}{n+1}\right)$$

$$= \left(1 - \frac{1}{2}\right) + \left(\frac{1}{2} - \frac{1}{3}\right) + \ldots + \left(\frac{1}{N} - \frac{1}{N+1}\right) = 1 - \frac{1}{N+1},$$

and $A_N \to 1$ as $N \to \infty$.

(ii) The *alternating series* $\sum_{n=1}^{\infty} \frac{(-1)^n}{n}$ is *conditionally convergent* since although it converges, the series

$$\sum_{n=1}^{\infty} \left|\frac{(-1)^n}{n}\right| = \sum_{n=1}^{\infty} \frac{1}{n} \quad \text{is divergent.}$$

(iii) The series $\sum_{n=0}^{\infty} \frac{z^n}{n!}$ is *absolutely convergent* for all z since $\sum_{n=0}^{\infty} \left|\frac{z^n}{n!}\right|$ converges for all z.

(iv) The series $1 + \frac{1}{3} - \frac{1}{2} + \frac{1}{5} + \frac{1}{7} - \frac{1}{4} + \frac{1}{9} + \frac{1}{11} - \frac{1}{6} + \ldots$ is a *rearrangement* of the *harmonic alternating series*

$$\sum_{n=1}^{\infty} \frac{(-1)^n}{n}.$$

It can be shown that the rearranged series does *not* converge to the same sum as the original series.

(Note that in several examples the series is summed from $n = 1$. Whether one sums from 0 or 1 does not affect the definition of convergence.)

Let $\sum_{n=0}^{\infty} a_n$, $\sum_{n=0}^{\infty} b_n$ be given series. The series $\sum_{n=0}^{\infty} c_n$ defined by

$$c_n = a_0 b_n + a_1 b_{n-1} + \ldots + a_n b_0,$$

is called the **Cauchy product** of the series Σa_n and Σb_n.

The series $\sum_{n=0}^{\infty} a_n$, of real or complex numbers, is said to have a $(C, 1)$ sum (a **Cesàro sum**), A, if $t_n \to A$ as $n \to \infty$, where

$$t_n = \frac{A_0 + A_1 + \ldots + A_n}{n+1} \quad (A_i \text{ are the partial sums of } \Sigma a_n).$$

Note. (i) The Cauchy product of two convergent series is not necessarily convergent.

(ii) If Σa_n is convergent, then its sum and its $(C, 1)$ sum agree.

Example. The series $\sum_{n=0}^{\infty} (-1)^n = 1-1+1-1+1-1+\ldots$ is not convergent, but has $(C, 1)$ sum $\frac{1}{2}$, for $t_{2n} = (n+1)/(2n+1)$, $t_{2n+1} = (n+1)/(2n+2)$, and $t_{2n} \to \frac{1}{2}$, $t_{2n+1} \to \frac{1}{2}$ as $n \to \infty$.

Power series

An infinite series of the form

$$a_0 + a_1(z-z_0) + a_2(z-z_0)^2 + \ldots + a_n(z-z_0)^n + \ldots$$

where the coefficients $a_0, a_1, a_2, \ldots$ are all constants is called a **power series** in $(z-z_0)$. (Here z, the constant z_0, and $a_0, a_1, \ldots$ may be real or complex numbers – if all are real then the power series is a *real power series*, otherwise it is a *complex* one. Since the former is a particular case of the latter we shall assume below that the power series is complex.)

With every power series one can associate a circle in $\mathbb{R}^2$, called the **circle of convergence**, such that for every z interior to the circle the series converges absolutely and for every z outside the circle the series diverges. The centre of the circle is at z_0 and its radius is known as the **radius of convergence** of the power series. (The radius of convergence may be zero or infinite.)

(If the power series is real, then we speak of the **interval of convergence**.)

If the power series converges for each z in some neighbourhood N of z_0, then we can define a function f on N by

$$f(z) = \sum_{n=0}^{\infty} a_n(z-z_0)^n.$$

The right-hand side of this relation is said to be a **power series expansion** for f about z_0 (see Taylor (p. 212), Maclaurin (p. 208) and Laurent (p. 207) series).

Examples. (i) The series $\sum_{n=0}^{\infty} \frac{z^n}{n!}$ has an infinite *radius of convergence.*

(ii) The series $\sum_{n=1}^{\infty} \frac{z^n}{n^2}$ has *radius of convergence* 1. Since $\sum_{n=1}^{\infty} \frac{1}{n^2}$ is convergent, the series is convergent everywhere on the *circle of convergence.*

Series of functions

The notion of a power series can be generalised to the case when each term of the series is a function – not necessarily a power – of a real or complex variable.

Given a sequence of functions (f_n) (p. 120) defined on a set T (in the case of functions of a complex variable, T will be a subset of $\mathbb{R}^2$), we can define a second sequence (s_n) by

$$s_n(z) = \sum_{i=0}^{n} f_i(z) \quad (z \in T, n \in \mathbb{N}).$$

If the sequence (s_n) converges uniformly (p. 120) on T to a function f, then we say that the series $\sum_{n=0}^{\infty} f_n(z)$ **converges uniformly** on T, and we write

$$\sum_{n=0}^{\infty} f_n(z) = f(z) \quad \text{(uniformly on } T\text{)}.$$

Note. (i) It can be shown that a power series converges uniformly on every compact subset (p. 110) interior to its circle of convergence.

(ii) Examples of series of functions can be found in § 38.

Double series

Consider the doubly infinite matrix of real or complex numbers

$$\begin{pmatrix} a_{11} & a_{12} & \dots & a_{1n} & \dots \\ a_{21} & a_{22} & \dots & a_{2n} & \dots \\ \vdots & \vdots & & \vdots & \\ a_{m1} & a_{m2} & \dots & a_{mn} & \dots \\ \vdots & \vdots & & \vdots & \end{pmatrix}$$

and suppose that for each fixed m the infinite series formed by the elements in the mth row, i.e. $a_{m1} + a_{m2} + a_{m3} + \dots$, converges to R_m, say. If the series $\sum_{m=1}^{\infty} R_m$ converges to R, say, then we say that the array of numbers is **convergent by rows** and that R is the **sum by rows** of the **double series**

$$\sum_{m=1}^{\infty}\left\{\sum_{n=1}^{\infty} a_{mn}\right\}.$$

In a similar fashion we can define (provided everything converges) the **column sum**

$$C = \sum_{n=1}^{\infty} C_n = \sum_{n=1}^{\infty}\left\{\sum_{m=1}^{\infty} a_{mn}\right\}.$$

Other sums can also be defined, such as the **square sum**, which is the limit of the sequence (S_i) where S_i denotes the finite sum of all terms a_{jk} such that $1 \leqslant j \leqslant i$, $1 \leqslant k \leqslant i$.

Note. It can be proved that, if the array of moduli $(|a_{mn}|)$ has a sum by one method, then the array (a_{mn}) has a sum by every method and that all methods give the same sum, known as the **sum of the double series** Σa_{mn}.

Example. Consider

$$R = \sum_{m=1}^{\infty}\sum_{n=1}^{\infty} \frac{(-1)^n}{(m+n^2)(m+n^2-1)}$$

(the *sum by rows* of the array defined by $a_{mn} = (-1)^n/(m+n^2)(m+n^2-1)$) and

$$C = \sum_{n=1}^{\infty}(-1)^n \sum_{m=1}^{\infty} \frac{1}{(m+n^2)(m+n^2-1)}$$

(the *sum by columns* of the array (a_{mn})).

Now $$\sum_{m=1}^{\infty} \frac{1}{(m+n^2)(m+n^2-1)} = \sum_{m=1}^{\infty}\left(\frac{1}{(m+n^2-1)} - \frac{1}{m+n^2}\right)$$

$$= \frac{1}{n^2} - \frac{1}{n^2+1} + \frac{1}{n^2+1} - \frac{1}{n^2+2} + \frac{1}{n^2+2} - \ldots = \frac{1}{n^2}.$$

Hence $$C = \sum_{n=1}^{\infty} \frac{(-1)^n}{n^2}$$

and C is *absolutely convergent*. It follows, then, from the result quoted in the above note, that R is convergent and converges to the same sum as C, namely

$$\sum_{n=1}^{\infty} \frac{(-1)^n}{n^2} = \frac{-\pi^2}{12}.$$

Infinite products

Analogous to the construction of infinite series is that of **infinite products.**

Given a sequence (a_n) of non-zero real or complex numbers we form a second sequence (P_n) whose terms are the **partial products**

$$P_n = \prod_{k=0}^{n} a_k = a_0 a_1 \dots a_n.$$

If the partial products P_n tend to the limit P as $n \to \infty$, then we say that P is the **value** of the **infinite product** $\prod_{n=0}^{\infty} a_n$.

If the sequence of partial products (P_n) is not convergent, then the infinite product is said to be **divergent**. It is convenient to distinguish those products for which (P_n) converges to a non-zero limit from those for which $P_n \to 0$ as $n \to \infty$, and for that reason it is usual to describe products with the latter property as **diverging to zero.**

Again, as in the case of infinite series one can consider products where the factors are functions of a real or complex variable, and can introduce the notions of *absolute* and *uniform convergence.*

Examples. (i) $\prod_{n=1}^{\infty}\left(1+\frac{1}{n}\right)$ is *divergent*, for

$$P_n = \prod_{k=1}^{n}\left(1+\frac{1}{k}\right) = \left(1+\frac{1}{1}\right)\left(1+\frac{1}{2}\right)\left(1+\frac{1}{3}\right)\dots\left(1+\frac{1}{n}\right)$$

$$= \frac{(2)(3)(4)\dots(n+1)}{n!} = n+1,$$

and as $n \to \infty$, $P_n \to \infty$.

(ii) $\prod_{n=2}^{\infty}\left(1-\frac{1}{n^2}\right)$ is *convergent*, for

$$P_n = \prod_{k=2}^{n}\left(1-\frac{1}{k^2}\right) = \left(1+\frac{1}{2}\right)\left(1-\frac{1}{2}\right)\left(1+\frac{1}{3}\right)\left(1-\frac{1}{3}\right)\dots\left(1+\frac{1}{n}\right)\left(1-\frac{1}{n}\right)$$

$$= \frac{n+1}{2}\left(1-\frac{1}{2}\right)\left(1-\frac{1}{3}\right)\dots\left(1-\frac{1}{n}\right)$$

$$= \frac{n+1}{2}\,\frac{(1.2.\dots.(n-1))}{n!} = \frac{n+1}{2n},$$

and as $n \to \infty$, $P_n \to \frac{1}{2}$.

See also: Abel's Limit Theorem (p. 198); Abel's test for convergence (p. 198); Cauchy's condensation test (p. 199); Cauchy's root test (p. 200); D'Alembert's test (p. 201); Dirichlet's tests (p. 202); Leibniz test (p. 207); Maclaurin series (p. 208); Raabe's test (p. 209); Ratio test (p. 209); Taylor series (p. 212); Wallis's formulae (p. 213); Weierstrass's *M*-test (p. 213).

29 *Improper integrals*

In § 27 we defined the integrals of functions which were defined and bounded on a finite interval. We now show how the definitions can be extended to cover *infinite integrals* (or *improper integrals of the first kind*) and integrals in which the interval remains finite but the integrand is unbounded (sometimes called *improper integrals of the second kind*). **The definitions given are for improper Riemann–Stieltjes integrals having real-valued integrand and integrator. The corresponding definitions for the Riemann integral are obtained by setting $\alpha(x) = x$ throughout.**

Assume that f is Riemann integrable with respect to α on all intervals $\leqslant a, b \geqslant$ for which $b \geqslant a$ and define a function I on $\leqslant a, \infty >$ by

$$I(b) = \int_a^b f(x)\,\mathrm{d}\alpha(x) \quad (a \leqslant b).$$

If the function so defined possesses a limit as $b \to \infty$, i.e. if $\lim_{b \to \infty} I(b) = l$ say, then we say that the **infinite integral** of f over $\leqslant a, \infty >$, denoted by

$$\int_a^\infty f(x)\,\mathrm{d}\alpha(x),$$

converges, otherwise it is said to **diverge**. If $I(b)$ converges to l as $b \to \infty$, then we say that l is the **value of the integral** and write

$$l = \int_a^\infty f(x)\,\mathrm{d}\alpha(x) = \int_a^\infty f\,\mathrm{d}\alpha.$$

Note the similarity between this definition and that of the sum of an infinite series (p. 140). Here the function values $I(b)$ play the same role as did the partial sums in the earlier definition.

Integrals over intervals of the type $< -\infty, a \geqslant$ are defined similarly. If both

$$\int_a^\infty f\,\mathrm{d}\alpha \quad \text{and} \quad \int_{-\infty}^a f\,\mathrm{d}\alpha$$

converge, then we say that $\int_{-\infty}^\infty f\,\mathrm{d}\alpha$ is convergent and define its value to be

$$\int_{-\infty}^a f\,\mathrm{d}\alpha + \int_a^\infty f\,\mathrm{d}\alpha.$$

Note. (i) It can be shown that the value of the above integral is independent of the choice of a.

(ii) Even in the case of the Riemann integral, the existence of

$$\lim_{b\to\infty}\int_{-b}^{b} f\,d\alpha$$

is a necessary, *but not sufficient*, condition for the convergence of the integral

$$\int_{-\infty}^{\infty} f\,d\alpha.$$

In cases where the former limit exists, e.g. when $f(x) = \alpha(x) = x$, but the latter integral does not converge, the limit is known as the **Cauchy principal value** of the integral.

(iii) Every convergent infinite series of real terms (p. 140) can be expressed as a convergent infinite integral. We have (with a definition of f analogous to that given in note (vi) on p. 138)

$$\sum_{i=1}^{\infty} a_i = \int_0^{\infty} f(x)\,d[x].$$

(iv) It can be shown (cf. the General Principle of Convergence, p. 115) that $\int_a^{\infty} f(x)\,dx$ converges if and only if for all $\epsilon > 0$ there exists x_0 such that

$$\left|\int_b^c f(x)\,dx\right| < \epsilon$$

where b and c are *any* numbers greater than x_0.

Examples. (i) $\int_0^{\infty} \sin x\,dx$ *diverges*, since $\int_0^b \sin x\,dx = 1-\cos b$ and $1-\cos b$ does not converge to a limit as $b\to\infty$.

(ii) $\int_{\frac{1}{2}\pi}^{\infty} \frac{\sin x}{x}\,dx$ *converges*. For suppose $\frac{1}{2}\pi < b < c$, then

$$\int_b^c \frac{\sin x}{x}\,dx = \left[\frac{-\cos x}{x}\right]_b^c - \int_b^c \frac{\cos x}{x^2}\,dx \quad \text{(see note (vii) p. 138),}$$

and

$$\left|\int_b^c \frac{\sin x}{x}\,dx\right| < \frac{1}{b}+\frac{1}{c}+\int_b^c \frac{dx}{x^2} = \frac{2}{b}.$$

Hence, using the result quoted in note (iv) above, the improper integral converges.

If α is monotonic increasing in $\leqslant a,\infty >$ and f is Riemann integrable with respect to α on all intervals $\leqslant a, b\geqslant$ $(b \geqslant a)$, then we say that the integral

$$\int_a^{\infty} f\,d\alpha$$

converges **absolutely** if
$$\int_a^\infty |f|\, d\alpha$$
converges. An integral which is convergent but not absolutely convergent is said to be **conditionally convergent.**

Let $f(x, y)$ be a real-valued function of two real variables defined on a subset of $\mathbb{R} \times \mathbb{R}$ of the form $\leqslant a, \infty > \times S$ where $S \subset \mathbb{R}$, such that
$$\phi(y) = \int_a^\infty f(x, y)\, d\alpha(x)$$
is convergent for all $y \in S$. The integral is said to be **uniformly convergent** on S if, given $\epsilon > 0$, there exists a number $N \in \mathbb{R}$ (depending on ϵ, but *not* on y), such that $x > N$ implies
$$\left|\phi(y) - \int_a^x f(x, y)\, d\alpha(x)\right| < \epsilon \quad \text{for all} \quad y \in S.$$

Example. $\displaystyle\int_{\frac{1}{2}\pi}^\infty \frac{\sin x}{x}\, dx$ is *conditionally convergent*. For, if $n \in \mathbb{N} - \{0\}$,
$$\int_{\frac{1}{2}\pi}^{n\pi} \left|\frac{\sin x}{x}\right| dx = \int_{\frac{1}{2}\pi}^{\pi} \frac{\sin x}{x}\, dx - \int_{\pi}^{2\pi} \frac{\sin x}{x}\, dx + \dots + (-1)^{n-1} \int_{(n-1)\pi}^{n\pi} \frac{\sin x}{x}\, dx$$
$$= \int_{\frac{1}{2}\pi}^{\pi} \frac{\sin x}{x}\, dx + \int_0^\pi \sin x \left(\frac{1}{x+\pi} + \dots + \frac{1}{x+(n-1)\pi}\right) dx$$
$$> \int_{\frac{1}{2}\pi}^{\pi} \frac{\sin x}{x}\, dx + \left(\frac{1}{2\pi} + \dots + \frac{1}{n\pi}\right) \int_0^\pi \sin x\, dx$$
$$= \int_{\frac{1}{2}\pi}^{\pi} \frac{\sin x}{x}\, dx + \frac{2}{\pi}\left(\frac{1}{2} + \frac{1}{3} + \dots + \frac{1}{n}\right),$$
and as $n \to \infty$,
$$\int_{\frac{1}{2}\pi}^{n\pi} \left|\frac{\sin x}{x}\right| dx \to \infty.$$

Unbounded integrands

If f is defined on the half-open finite interval $< a, b \geqslant$ and is Riemann integrable with respect to α on all intervals $\leqslant c, b \geqslant$ where $a < c \leqslant b$, we can define a function I on $< a, b \geqslant$ by
$$I(c) = \int_c^b f\, d\alpha \quad (a < c \leqslant b).$$

If the function I possesses a limit as $c \to a^+$, then we say that the integral
$$\int_{a+}^b f\, d\alpha$$

converges and has **value** equal to $l = \lim_{c \to a+} I(c)$. We write

$$\int_{a+}^{b} f\,d\alpha = l.$$

In the case of the Riemann integral it is permissible and customary to write

$$\int_a^b f(x)\,dx \quad \text{or} \quad \int_a^b f.$$

Similar definitions can be provided for integrals of the form

$$\int_a^{b-} f\,d\alpha, \quad \int_{a+}^{\infty} f\,d\alpha \left(= \int_{a+}^{b} + \int_b^{\infty} \right) \quad \text{and so on.}$$

Definitions analogous to those given on pp. 147–8 for *absolutely* and *uniformly convergent infinite integrals* can also be framed.

An improper integral of particular interest is the integral

$$\int_{0+}^{\infty} e^{-x} x^{p-1}\,dx,$$

which converges for $p > 0$. The value of the integral is denoted by $\Gamma(p)$ and the function Γ is known as the **gamma function.** The function satisfies the relation

$$\Gamma(p) = (p-1)\,\Gamma(p-1) \quad (p > 1)$$

and, in particular, when $n \in \mathbb{N}$ we have

$$\Gamma(n+1) = n(n-1)\,\ldots\,2.1;$$

a product known as **factorial** n and denoted by $n!$

Examples. (i) $\int_0^1 \frac{1}{\sqrt{x}}\,dx$ *converges*, since, if $0 < a < 1$,

$$I(a) = \int_a^1 \frac{1}{\sqrt{x}}\,dx = [2\sqrt{x}]_a^1 = 2 - 2\sqrt{a}$$

and $I(a) \to 2$ as $a \to 0$.

(ii) $\int_0^{\frac{1}{2}\pi} \frac{\sin x}{x}\,dx$ can be shown to *converge*. (Note that (ii) presents a different problem from (i), for it is possible to extend the domain $(\mathbb{R} - \{0\})$ of the function $f : x \mapsto \sin x/x$ in such a way that the extension is continuous at $x = 0$ (set $f(0) = 1$). It is not possible, however, to do this in the case of the function $g : x \mapsto 1/\sqrt{x}$.)

30 Curves and arc length

A continuous, vector-valued function $\boldsymbol{\alpha}$ mapping a closed interval $\leqslant a, b \geqslant$ of $\mathbb{R}$ into $\mathbb{R}^k$ is called a **curve** or **path** in $\mathbb{R}^k$. If $\boldsymbol{\alpha}$ is an injection (p. 14), then $\boldsymbol{\alpha}$ is called an **arc.** If $\boldsymbol{\alpha}(a) = \boldsymbol{\alpha}(b)$, then $\boldsymbol{\alpha}$ is a **closed curve**; if, in addition, $\boldsymbol{\alpha}(t_1) \neq \boldsymbol{\alpha}(t_2)$ for any other distinct points t_1, t_2 of $\leqslant a, b \geqslant$, then $\boldsymbol{\alpha}$ is called a **simple closed curve** or a **Jordan curve.** The **reverse** of the curve $\boldsymbol{\alpha}$ is the function defined on $\leqslant a, b \geqslant$ by $t \mapsto \boldsymbol{\alpha}((b+a)-t)$.

Note. (i) It is now becoming increasingly common to define a curve to be a *mapping* and not the *image* of $\leqslant a, b \geqslant$ under a mapping. This usage is, however, by no means universal and care must be taken, therefore, to note which definition an author is adopting. Using the definition given above, the same point set in $\mathbb{R}^k$ can be associated with different curves, whereas if the alternative definition is used the same curve can be described by several functions. In this latter case, functions which give rise to the same curve are said to be **equivalent – properly equivalent** if they are both associated with the same 'direction' on the curve, and **improperly equivalent** otherwise. (Thus, for example, the functions $t \mapsto (t, 2-t)$ and $t \mapsto (t^2, 2-t^2)$ with domain $\leqslant 0, 1 \geqslant$ are properly equivalent, whereas either of these is improperly equivalent to $t \mapsto (1-t, 1+t)$.) Since Jordan curves have no 'physically obvious' end-points this gives rise to certain complications which are resolved by saying that two functions defining Jordan curves are *equivalent* (properly or improperly as before) if and only if they give rise to the same point set.

(ii) An alternative definition of an *arc* is a curve which is not closed. With this usage an arc as defined above is referred to as a *simple* or *Jordan arc.*

(iii) Some authors differentiate between the use of *path* and *curve* by using one to indicate a point set without any associated direction and the other to indicate the set together with an associated direction, i.e. a *directed path* or an *oriented curve.* (Regrettably there is no unanimity over which term denotes the point set without direction.)

The simplest type of curve is that given by a formula of the type

$$\boldsymbol{\alpha}(t) = t\mathbf{p}+(1-t)\mathbf{q} \quad (t \in \leqslant 0, 1 \geqslant),$$

where $\mathbf{p}$ and $\mathbf{q}$ are points in $\mathbb{R}^k$. $\boldsymbol{\alpha}$ is then said to be a **straight line segment** from $\mathbf{p}$ to $\mathbf{q}$ and we define the **length** of the segment to be $|\mathbf{p}-\mathbf{q}| \,(=\|\mathbf{p}-\mathbf{q}\|$, see p. 169).

If, now, $\boldsymbol{\alpha}$ is any curve in $\mathbb{R}^k$ and $P = \{a_0, \ldots, a_n\}$ is a partition (p. 118) of the interval $\leqslant a, b \geqslant$, then the points of the set

$$\{\boldsymbol{\alpha}(a_0), \boldsymbol{\alpha}(a_1), \ldots, \boldsymbol{\alpha}(a_n)\}$$

are called the *vertices of the inscribed polygon* $\Pi(P)$ determined by P.

The *length of the polygon* is then taken to be the sum of the lengths of the straight line segments joining neighbouring vertices, i.e. to be

$$\sum_{i=1}^{n} |\alpha(a_i) - \alpha(a_{i-1})|.$$

If there exists a positive number M such that the length of $\Pi(P)$ is less than M for all possible partitions P of $\leqslant a, b \geqslant$, then α is said to be a **rectifiable** curve. If so, the **length** of α is defined to be the least upper bound of the lengths of inscribed polygons taken over all possible partitions.

Note. (i) α is rectifiable if and only if it is of *bounded variation*, i.e. if each of its k component functions $t \mapsto \pi^n(\alpha(t))$, $n = 1, 2, \ldots, k$, are of bounded variation (p. 118) over $\leqslant a, b \geqslant$.

(ii) It can be shown that if α' exists and is continuous on $\leqslant a, b \geqslant$, then α is rectifiable and has length.

$$\int_a^b |\alpha'(t)| \, dt.$$

A curve $\alpha : \leqslant a, b \geqslant \to \mathbb{R}^k$ is said to be **piecewise smooth** if each component $\alpha_1, \ldots, \alpha_k$ of α has a bounded derivative α_i' $(i = 1, \ldots, k)$ which is continuous everywhere in $\leqslant a, b \geqslant$ except (possibly) at a finite number of points at which left- and right-handed derivatives (p. 122) exist.

Note. (i) Every piecewise smooth curve is rectifiable.

(ii) Every rectifiable curve can be approximated by piecewise smooth curves (in particular, by inscribed polygons).

Examples. (i) Consider the *curve* $\alpha : t \mapsto (\cos t, \sin t)$ defined on the interval $\leqslant 0, 2\pi \geqslant$. Then, since $\alpha(0) = \alpha(2\pi)$ and $\alpha(t_1) \neq \alpha(t_2)$ for any other distinct points t_1 and t_2 in $\leqslant 0, 2\pi \geqslant$, α is a *Jordan curve*. The *reverse* of α is the curve defined on $\leqslant 0, 2\pi \geqslant$ by $t \mapsto (\cos(2\pi - t), \sin(2\pi - t))$, i.e. $t \mapsto (\cos t, -\sin t)$. α' exists and is continuous on $\leqslant 0, 2\pi \geqslant$ and so it follows (note (ii) above) that α is *rectifiable* and has *length*

$$\int_0^{2\pi} |(-\sin t, \cos t)| \, dt = \int_0^{2\pi} \sqrt{(\sin^2 t + \cos^2 t)} \, dt = \int_0^{2\pi} dt = 2\pi.$$

(ii) Consider the curve

$$\beta : t \mapsto \begin{cases} (t, t \sin 1/t), & t \neq 0, \\ (0, 0), & t = 0. \end{cases}$$

β is not rectifiable on $\leqslant 0, 1 \geqslant$ since it is not of bounded variation on that interval (see p. 119).

Connectedness and regions

A set X in $\mathbb{R}^k$ is said to be **arcwise connected** if for every pair of points $\mathbf{p}$, $\mathbf{q}$ in X there exists a curve $\boldsymbol{\alpha}: \leqslant a, b \geqslant \to X$ such that

$$\boldsymbol{\alpha}(a) = \mathbf{p}, \quad \boldsymbol{\alpha}(b) = \mathbf{q}.$$

More intuitively: if every pair of points in X can be joined by a curve lying wholly in X.

If $\boldsymbol{\alpha}$ can always be chosen to be a straight-line segment (p. 150), then X is said to be **convex.**

Note. (i) Every arcwise connected set is connected in the sense of the definition on p. 110.

(ii) The converse of (i) is not true (see example below). It is true, however, that every *open* connected set is arcwise connected.

Example. The set $S = A \cup B$ where

$$A = \{(x, y) \mid x \in <0, 1\geqslant, y = \sin 1/x\},$$

$$B = \{(x, y) \mid x \in \leqslant -1, 0\geqslant, y = 0\},$$

is connected (p. 110) but *not* arcwise connected since no point in A can be joined to the point $(-1, 0) \in B$ by a path in S (S is not locally connected p. 110).

A subset of $\mathbb{R}^k$ is called a **region** if it is the union of an open connected set with some, none or all of its frontier points (p. 102). If none of the frontier points is included, the region is known as an **open region,** if all are included, as a **closed region.**

Given a Jordan curve $\boldsymbol{\alpha}$ in $\mathbb{R}^2$, the *Jordan Curve Theorem* states that if C is the image of $\boldsymbol{\alpha}$, then $\mathbb{R}^2 - C = A \cup B$ where A and B are two open regions exactly one of which is bounded. Moreover,

$$\mathrm{Fr}(A) = \mathrm{Fr}(B) = C \quad \text{(p. 102)}.$$

The bounded component of $\mathbb{R}^2 - C$ is called the **interior of $\boldsymbol{\alpha}$** (or the **interior of** C) and the unbounded component the **exterior.**

Subsets of the interior are said to be **inside** C and those of the exterior to be **outside** C.

31 Functions of a complex variable

A function f having domain D in the complex plane $\mathbb{R}^2$ (or E_2) (p. 68) and codomain either $\mathbb{R}$ or $\mathbb{R}^2$ is said to tend to a **limit** l as z tends to $z_0 \in \bar{D}$ if given any $\epsilon > 0$, we can find a δ such that

$$|f(z) - l| < \epsilon$$

whenever $z \in D$ and $0 < |z - z_0| < \delta$.

When this is the case we write $\lim_{z \to z_0} f(z) = l$.

If the codomain of f is $\mathbb{R}$, then f is called a *real-valued function of a complex variable*; if $\mathbb{R}^2$, then f is a *complex-valued function.*

f is said to be **continuous** at z_0 if the limit as z tends to z_0 exists and is equal to $f(z_0)$.

f is said to be **continuous** on D if it is continuous at every point of D.

f is said to be **uniformly continuous** on D if, given any $\epsilon > 0$, we can find a δ (depending only on ϵ) which will satisfy the requirements of the definition of continuity at every point of D.

Note. Those definitions are, of course, particular cases of the more general definitions contained on pp. 103 and 106. Here, the domain of the mapping is a subset of the complex plane, the metric $d(z_1, z_2)$ is $|z_1 - z_2|$, and the codomain is the real line or the complex plane, again with the usual metric.

f is said to be **bounded** on D if there exists a positive constant K such that $|f(z)| < K$ for all $z \in D$.

If f is a complex-valued function defined on an open set S in $\mathbb{R}^2$ and $z_0 \in S$, then we say that f is **differentiable** at z_0 if the limit

$$\lim_{z \to z_0} \frac{f(z) - f(z_0)}{z - z_0}$$

exists, and the value of the limit is defined to be the **derivative** of f at z_0, denoted by $f'(z_0)$. (Compare the definition on p. 122.)

Examples. (i) The *real-valued function*

$$x + iy \mapsto \begin{cases} \dfrac{xy}{x^2 + y^2} & (x + iy \neq 0), \\ 0, & (x + iy = 0), \end{cases}$$

is *not* continuous at $z = 0$ (see example (i) p. 130).

(ii) Let f be the *complex-valued function* $z \mapsto z^n$. Then

$$\lim_{z \to z_0} \frac{f(z)-f(z_0)}{z-z_0}$$

exists and is equal to nz_0^{n-1}, i.e. f is *differentiable* at z_0 and the *derivative* of f at z_0, $f'(z_0)$, is nz_0^{n-1}.

(iii) The complex-valued function $z \mapsto |z|^2$ is *differentiable* at the origin, but nowhere else. For

$$\frac{|z|^2-|z_0|^2}{z-z_0} = \frac{z\bar{z}-z_0\bar{z}_0}{z-z_0} = \bar{z}+z_0\frac{\bar{z}-\bar{z}_0}{z-z_0} = \bar{z}+z_0(\cos 2\theta - i\sin 2\theta),$$

where $\theta = \arg(z-z_0)$. It is obvious that this last expression does not tend to a unique limit as $z \to z_0$ except in the case $z_0 = 0$, when it tends to 0, i.e. $f'(0) = 0$.

Note. If f is a complex-valued function of a complex variable then each value of f can be written in the form

$$f(z) = u(z)+iv(z)$$

where u and v are real-valued functions. One can consider u and v to be real-valued functions of two real variables, i.e. write

$$f(z)\ (=f(x+iy)) = u(x, y)+iv(x, y)$$

(f is thought of as a vector-valued function $\mathbf{f}: \mathbb{R}^2 \to \mathbb{R}^2$ with components u and v).

Further consideration of this presentation leads us to obtain conditions for the differentiability of z in terms of the partial derivatives of u and v, see Cauchy–Riemann equations (p. 200).

If a complex-valued function f is differentiable at every point of some open set S of $\mathbb{R}^2$, then f is said to be **analytic** on S.

A function is said to be **analytic at a point** z_0 if there is a neighbourhood of z_0 on which it is analytic.

Note. (i) It can be shown that if f' exists on S it is necessarily continuous on S. This latter (superfluous) condition is, however, sometimes included for pedagogical reasons in the definition of an analytic function.

(ii) If a function f is analytic at a point z_0 then there exists a convergent power series (p. 142) which represents the function f in a neighbourhood of z_0 and conversely. For this reason the existence of a power series representation is often taken as a starting point in the definition of an analytic function.

(iii) Some authors weaken the definition of an analytic function on a domain S by asking only that the function should be analytic (in the above sense) at all but a finite number of points of S. A function which is analytic in the stronger sense would be described by them as being **regular** on S.

Points at which a function is not analytic are called the **singular points** or **singularities** of the function.

A point z_0 is said to be an **isolated singularity** of f if f is *not* analytic at z_0 but is analytic elsewhere throughout some neighbourhood of z_0. (We do not insist that z_0 should be in the domain of f.)

A set comprising a neighbourhood of a point minus the point itself is often called a **punctured** or **deleted neighbourhood.**

It can be shown (see Laurent's Theorem, p. 207) that if z_0 is an isolated singularity, then in the punctured neighbourhood of z_0 in which f is analytic, f can be expanded in the form (i.e. can be represented by the series)

$$f(z) = \sum_{n=0}^{\infty} a_n(z-z_0)^n + \sum_{n=1}^{\infty} b_n(z-z_0)^{-n}.$$

This expansion is known as the **Laurent expansion.**

If $b_n = 0$ for $n = 1, 2, \ldots$, the point z_0 is called a **removable singularity** (since the function becomes analytic if we set $f(z_0) = a_0$).

If $b_m \neq 0$ for some m but $b_n = 0$ for all $n > m$, then z_0 is said to be a **pole** of **order** m.

A pole of order 1 is said to be a **simple pole.**

If $b_n \neq 0$ for infinitely many values of n, then z_0 is said to be an **essential singularity**. (A singularity is also said to be **essential** if it is not an isolated one.)

If z_0 is an isolated singular point of f, then the coefficient b_1 in the Laurent expansion for f is called the **residue** of f at z_0 and is denoted by $\operatorname{Res}_{z=z_0} f(z)$.

Note. It can be shown that if f has a simple pole at z_0, then

$$\operatorname{Res}_{z=z_0} f(z) = \lim_{z\to z_0} (z-z_0)f(z),$$

that if f has a double pole at z_0, then

$$\operatorname{Res}_{z=z_0} f(z) = g'(z_0) \quad \text{where} \quad g(z) = (z-z_0)^2 f(z),$$

and $g(z_0)$ is defined to be $\lim_{z\to z_0} (z-z_0)^2 f(z)$, and so on.

If f is analytic at z_0, then the Laurent expansion takes the form

$$f(z) = \sum_{n=0}^{\infty} a_n(z-z_0)^n \quad \text{(cf. Taylor expansion p. 212).}$$

If $f(z_0) = 0$, then $a_0 = 0$ and we say f has a **zero** at z_0. If $a_n = 0$ for all n, then we say that f is **identically zero** in the neighbourhood. If $a_m \neq 0$ and $a_n = 0$ for all $n < m$, then we say that f has a **zero of order** m at z_0.

Examples. (i) The function $f: z \mapsto z^n$ is *analytic* on every open region of $\mathbb{R}^2$ (see example (ii) p. 154).

(ii) The function $g: z \mapsto \sin z/z$ if $z \neq 0$, $g(0) = 0$, is *analytic* everywhere except at 0.

The *Laurent expansion* about 0 for g is

$$1 - \frac{z^2}{3!} + \frac{z^4}{5!} - \dots.$$

We note then that 0 is a ***removable singularity*** of g, and that if h is defined by

$$h(z) = 1 - \frac{z^2}{3!} + \frac{z^4}{5!} - \dots,$$

then h is *analytic* everywhere.

(iii) Let f be the function $z \mapsto \cos z/z^4$ $(z \neq 0)$. Then the *Laurent expansion* for f about 0 is

$$\frac{1}{z^4} - \frac{1}{2z^2} + \frac{1}{4!} - \frac{z^2}{6!} + \dots,$$

and $f(z)$ has a **pole of order** 4 at $z = 0$. The *residue* of $f(z)$ at the origin is 0.

(iv) The function $g: z \mapsto \sec(1/z)$ has *isolated singularities* (in fact, *simple poles*) at the points $z = 1/(n+\frac{1}{2})\pi$ where $n \in \mathbb{Z}$. It has an *essential singularity* at the origin since the singularity there is not isolated.

(v) The origin is an *essential singularity* of $f: z \mapsto e^{1/z}$ $(z \neq 0)$. In this case the point 0 is an *isolated singularity*, and the *Laurent expansion* of f about 0 is

$$1 + z^{-1} + \frac{1}{2!} z^{-2} + \dots + \frac{1}{n!} z^{-n} + \dots.$$

(vi) The function $g: z \mapsto z/(z^2+4)$ has *simple poles* at $z = \pm 2i$. The *residue* at $z = 2i$ is given by

$$\lim_{z \to 2i} (z - 2i) . \frac{z}{z^2+4} = \tfrac{1}{2}.$$

(vii) The function $f: z \mapsto 6(z-2)^3 + 5(z-2)^5$ is *analytic* everywhere and has a *zero of order* 3 at $z = 2$.

If f is a complex-valued function defined on an open region S in $\mathbb{R}^2$, then f is said to be **conformal** at $z_0 \in S$ if every pair of smooth arcs (pp. 150–1) Γ_1 and Γ_2 intersecting at z_0 is mapped by f onto a corresponding pair of smooth arcs in such a way that the angle from $f(\Gamma_1)$ to $f(\Gamma_2)$ at $f(z_0)$ is equal (modulo 2π) to that from Γ_1 to Γ_2 at z_0.

It can be shown that if f is analytic on S, then it is conformal at each point $z_0 \in S$ for which $f'(z_0) \neq 0$.

Contour integrals

Let f be a complex-valued function defined on the image of a curve $\alpha: \langle a, b \rangle \to \mathbb{C}$ (p. 150) let $P = \{a_0, \dots, a_n\}$ be a partition of $\langle a, b \rangle$.

If the sum

$$\sum_{i=1}^{n} f(z_i)\,(\alpha(a_i) - \alpha(a_{i-1})),$$

where z_i is some point $\boldsymbol{\alpha}(t_i)$ such that $a_{i-1} \leqslant t_i \leqslant a_i$, tends to a unique limit l as n tends to infinity and the greatest of the numbers $a_i - a_{i-1}$ tends to zero, then we say that the **contour integral** of f along $\boldsymbol{\alpha}$ exists and has value l. The contour integral is denoted by

$$\int_{\boldsymbol{\alpha}} f(z)\,\mathrm{d}z.$$

Note. (i) If $\operatorname{Im}(\boldsymbol{\alpha})$ is a segment of the real axis, then this definition reduces to that for the Riemann integral of $f(x)$ between $\boldsymbol{\alpha}(a)$ and $\boldsymbol{\alpha}(b)$. (See also note (iv) on p. 138.)

(ii) An alternative definition, making use of the Riemann–Stieltjes integral, is based on the fact that the definition on p. 137 can be extended without any other changes in the wording to cover the cases where f and α are complex-valued functions.

Now let $\boldsymbol{\alpha}$ be any curve $\leqslant a, b \geqslant \to \mathbb{R}^2$. Then $\boldsymbol{\alpha}$ can be expressed in terms of its components (α_1, α_2) (p. 151) and can be associated with the complex-valued function

$$z(t) = \alpha_1(t) + \mathrm{i}\alpha_2(t).$$

Given any complex-valued function of a complex variable, f say, defined on $\operatorname{Im}(\boldsymbol{\alpha})$ we define the **contour integral** of f along $\boldsymbol{\alpha}$, denoted by

$$\int_{\boldsymbol{\alpha}} f(z)\,\mathrm{d}z,$$

by

$$\int_{\boldsymbol{\alpha}} f(z)\,\mathrm{d}z = \int_a^b f(z(t))\,\mathrm{d}z(t),$$

whenever the complex Riemann–Stieltjes integral on the right exists.

(iii) Reversing the direction of a curve changes the sign of the integral.

(iv) The contour integral always exists if $\boldsymbol{\alpha}$ is rectifiable and f is continuous.

(v) If $\boldsymbol{\alpha}$ is piecewise smooth (p. 151) and the contour integral of f along $\boldsymbol{\alpha}$ exists, then

$$\int_{\boldsymbol{\alpha}} f\,\mathrm{d}z = \int_a^b f(z(t))\,z'(t)\,\mathrm{d}t \quad \text{(cf. note (vii) p. 138).}$$

Examples. (i) Establishing the existence of a contour integral by first principles is not an easy matter. However, only minor changes are necessary in the wording of the example on p. 136 to show that the integral of $f : z \mapsto z$ along a straight line segment joining $z = a$ to $z = b$ is $\frac{1}{2}(b^2 - a^2)$.

Otherwise, we can consider $\boldsymbol{\alpha} : t \mapsto a + (b-a)t$ defined on $\leqslant 0, 1 \geqslant$ and, using the result in note (v) above, write

$$\begin{aligned}\int_{\boldsymbol{\alpha}} z\,\mathrm{d}z &= \int_0^1 (a + (b-a)t)\,(b-a)\,\mathrm{d}t \\ &= (b-a)\,[at + (b-a)\,t^2/2]_0^1 \\ &= \frac{b^2 - a^2}{2}.\end{aligned}$$

(ii) Consider $\int_\alpha \frac{dz}{z}$ where α is defined by $t \mapsto (\cos t, \sin t)$ on $\leqslant 0, 2\pi \geqslant$.
Then $z(t) = \cos t + i \sin t = e^{it}$, and

$$\int_\alpha \frac{dz}{z} = \int_0^{2\pi} \frac{1}{e^{it}} ie^{it} dt = i \int_0^{2\pi} dt = 2\pi i.$$

If α is a rectifiable, closed curve (p. 151) in $\mathbb{R}^2$ and if z_0 is a point of $\mathbb{R}^2 - \text{Im}(\alpha)$, we define $W[\alpha : z_0]$, the **index** or **winding number** of the path α with respect to z_0, to be

$$\frac{1}{2\pi i} \int_\alpha \frac{1}{z - z_0} dz.$$

It can be shown that $W[\alpha : z_0] \in \mathbb{Z}$ and that if α is a rectifiable Jordan curve then $W[\alpha : z_0] = \pm 1$; all points 'inside' α (p. 152) will then have the same index and we can define the **orientation** of a Jordan curve α by saying that α is **positively oriented** if the index of every point inside α is $+1$ and **negatively oriented** if it is -1.

Examples. It follows from example (ii) above that $\alpha : t \mapsto (\cos t, \sin t)$ on $\leqslant 0, 2\pi \geqslant$ has *winding number* 1 with respect to the origin. α (which is a rectifiable Jordan curve, see example on p. 151) is therefore *positively oriented.*

$\beta : t \mapsto (\cos 2t, \sin 2t)$ on $\leqslant 0, 2\pi \geqslant$ is also rectifiable and $\text{Im}(\alpha) = \text{Im}(\beta)$. It is readily checked that β is not a Jordan curve, and that $W[\beta : 0] = 2$.

See also: Cauchy integral formula (p. 199); Cauchy Integral Theorem (p. 199); Cauchy Residue Theorem (p. 199); Cauchy–Riemann equations (p. 200); Laurent's Theorem (p. 207); Liouville's Theorem (p. 208); Rouché's Theorem (p. 210); Taylor series (p. 212).

32 Multiple integrals

We first generalise the concept of a Riemann integral over an interval $\leqslant a, b \geqslant \subset \mathbb{R}$ to that of an integral over an n-dimensional region of a particularly simple nature.

The Cartesian product of n closed intervals $\leqslant a_k, b_k \geqslant$ $(k = 1, 2, \ldots, n)$ is called an **n-dimensional closed interval** or an **n-cell** in $\mathbb{R}^n$ and we denote it by I^n.

The **content** or **measure** of I^n, denoted by $\mu(I^n)$, is defined to be the product

$$(b_1 - a_1)(b_2 - a_2) \ldots (b_n - a_n).$$

If P_k is a partition of $\leqslant a_k, b_k \geqslant$ $(k = 1, \ldots, n)$ then we define a partition of I^n by setting

$$P = P_1 \times P_2 \times \ldots \times P_n.$$

As before (p. 118), P' will be a *finer* partition than P if $P \subset P'$.

Let f be a real-valued function defined and bounded on a closed interval $I^n \subset \mathbb{R}^n$. Given any partition P of subintervals $I_1^n, \ldots, I_m^n$ of I^n we can form sums of the type

$$S(P, f) = \sum_{i=1}^{m} f(t_i)\mu(I_i^n),$$

where t_i is some element of I_i^n.

f is said to be **Riemann-integrable** on I^n whenever there exists $A \in \mathbb{R}$ satisfying: given any $\epsilon > 0$ there exists a partition P_ϵ of I such that whenever P is finer than P_ϵ it follows that

$$|S(P, f) - A| < \epsilon$$

for all possible sums $S(P, f)$.

If such a number exists it is uniquely determined, is known as the **value** of the **integral of f over I^n**, and is denoted by

$$\int_{I^n} f \, d\mathbf{x} \quad \text{or} \quad \int_{I^n} f(\mathbf{x}) \, d\mathbf{x} \quad \text{or} \quad \int_{I^n} f(x_1, \ldots, x_n) \, d(x_1, \ldots, x_n),$$

or (more particularly when $n = 2$ or 3)

$$\int_{a_1}^{b_1} \int_{a_2}^{b_2} \ldots \int_{a_n}^{b_n} f(x_1, \ldots, x_n) \, dx_1 \, dx_2 \ldots dx_n.$$

Note. As in the case of the integral of a function of a single variable, it is possible to frame this definition in terms of upper and lower Riemann sums.

We now seek to extend the concept of integration to regions of $\mathbb{R}^n$ other than n-dimensional closed intervals.

Let S be a subset of some interval I^n of $\mathbb{R}^n$. Given a partition P of I^n we define $\underline{J}(P, S)$ to be the sum of the measures (p. 159), of those subintervals of P which contain only interior points of S. Similarly, we define $\bar{J}(P, S)$ to be the sum of the measures of those subintervals of P which contain points of $S \cup \mathrm{Fr}(S)$ (p. 102). The glb of $\{\bar{J}(P, S)\}$ and the lub of $\{\underline{J}(P, S)\}$ taken over all partitions of I^n are known, respectively, as the **outer** and **inner Jordan content** of S. If these two contents are equal, then S is said to be **Jordan-measurable** with **Jordan content**, denoted by $c(S)$, equal to the common value.

In particular, if $n = 2$, then the set S is said to have **area** $c(S)$, if $n = 3$ then $c(S)$ is said to be the **volume** of S.

Let f be defined and bounded on a bounded, Jordan-measurable set $S \subset \mathbb{R}^n$ and let I^n be a closed interval of $\mathbb{R}^n$ containing S.

Define g on I^n by

$$g(\mathbf{x}) = \begin{cases} f(\mathbf{x}) & \text{if } \mathbf{x} \in S, \\ 0 & \text{if } \mathbf{x} \in I^n - S. \end{cases}$$

f is then said to be **Riemann integrable over** S whenever g is Riemann integrable over I^n and we define the value of the integral of f over S, written $\int_S f(\mathbf{x})\,d\mathbf{x}$, to be $\int_{I^n} g(\mathbf{x})\,d\mathbf{x}$.

It can be shown that this definition is independent of the choice of I^n.

Note. Multiple integrals are usually evaluated by '*repeated integration*', i.e. by making use of the result that if S is a closed region of $\mathbb{R}^2$ given by

$$S = \{(x, y) \mid a \leqslant x \leqslant b,\ \phi_1(x) \leqslant y \leqslant \phi_2(x)\},$$

where ϕ_1 and ϕ_2 are continuous functions such that, for all $x \in \langle a, b \rangle$, $\phi_1(x) \leqslant \phi_2(x)$, then

$$\int_S f(x, y)\,dx\,dy = \int_a^b \left[\int_{\phi_1(x)}^{\phi_2(x)} f(x, y)\,dy\right] dx$$

for all functions f which are Riemann-integrable on S. Similar statements can be made for functions of three or more variables.

It should be noted, however, that the existence of a repeated integral does not imply that of the associated multiple integral. For instance,

$$\iint_S \frac{x-y}{(x+y)^3}\,d(x, y)$$

where $S = \{(x, y) | 0 \leqslant x \leqslant 1, 0 \leqslant y \leqslant 1\}$ does not exist whereas

$$\int_0^1 \left[\int_0^1 \frac{x-y}{(x+y)^3}\,dx\right] dy = -\tfrac{1}{2} = -\int_0^1 \left[\int_0^1 \frac{x-y}{(x+y)^3}\,dy\right] dx.$$

Examples. (i) Consider the integral of $(x, y) \mapsto \sqrt{(a^2 - x^2 - y^2)}$ over the first quadrant of the circle $x^2 + y^2 = a^2$.

In this example the region of integration, S, is given by

$$S = \{(x, y) | 0 \leqslant x \leqslant a, 0 \leqslant y \leqslant \sqrt{(a^2 - x^2)}\},$$

and
$$\int_S \sqrt{(a^2 - x^2 - y^2)}\,dx\,dy = \int_0^a \left[\int_0^{\sqrt{(a^2-x^2)}} \sqrt{(a^2 - x^2 - y^2)}\,dy\right] dx$$
$$= \int_0^a \tfrac{1}{4}\pi(a^2 - x^2)\,dx$$
$$= \tfrac{1}{6}\pi a^3.$$

(ii) Consider the triple integral of $(x, y, z) \mapsto x^2y^2z$ over S, where S is the portion of the cone $x^2 + y^2 = xz$ lying between the planes $z = 0$ and $z = c$. Changing our coordinate system to that of *spherical polar coordinates* (p. 132), observing that

$$\frac{\partial(x, y, z)}{\partial(r, \theta, \phi)} = -r^2 \sin\phi,$$

and making use of the theorem on *change of variable* (p. 201) and the fact that $r^2 \sin\phi$ is non-negative for $0 \leqslant \phi \leqslant \pi$, we then have $\int_S x^2y^2z\,dx\,dy\,dz$

$$= 2\int_0^{\frac{1}{2}\pi} \left[\int_0^{\tan^{-1}(\cos\theta)} \left[\int_0^{c/\cos\phi} r^7 \sin^5\phi \cos\phi \cos^2\theta \sin^2\theta\,dr\right] d\phi\right] d\theta$$
$$= \frac{c^8}{4}\int_0^{\frac{1}{2}\pi} \cos^2\theta \sin^2\theta \left[\int_0^{\tan^{-1}(\cos\theta)} \sin^5\phi \sec^7\phi\,d\phi\right] d\theta$$
$$= \frac{c^8}{24}\int_0^{\frac{1}{2}\pi} \cos^8\theta \sin^2\theta\,d\theta = \frac{7\pi}{48}\left(\frac{c}{2}\right)^8.$$

See also: Change of variable (p. 201).

33 Logarithmic, exponential and trigonometric functions

The order in which the logarithmic and exponential functions are defined is a matter of choice; we give below two alternative sequences of definitions.

We define the **logarithm** of x (the **natural logarithm** of x), written $\log x$, to be

$$\int_1^x \frac{1}{t}\,dt \quad (x > 0).$$

Hence, log is a function with domain $\{x \in \mathbb{R} | x > 0\}$ and codomain $\mathbb{R}$. It can, moreover, be shown that log is bijective, and that it satisfies the relations:

(i) $\log ab = \log a + \log b \quad (a, b > 0)$;
(ii) $\log 1 = 0$;
(iii) $\log 1/a = -\log a \quad (a > 0)$;
(iv) $\log a^s = s \log a \quad (a > 0, s \in \mathbb{Q})$.

Since log is injective, the inverse image of 1 is uniquely defined. It is denoted by e, i.e. $\log^{-1} 1 = e$.

Since log is continuous and strictly increasing for $x > 0$, it follows from the Inverse Function Theorem (p. 205) that log possesses an inverse which is strictly increasing and continuous on $\mathbb{R}$. This function is denoted by exp and is referred to as the **exponential function.**

It follows, therefore, that exp is bijective, has domain $\mathbb{R}$ and codomain $\{x \in \mathbb{R} | x > 0\}$, and that:

(i) $\exp a . \exp b = \exp(a+b) \quad (\forall a, b \in \mathbb{R})$,
(ii) $\exp 0 = 1$,
(iii) $\exp(-a) = 1/\exp(a) \quad (\forall a \in \mathbb{R})$,
(iv) $\exp(x) = e^x \quad (\forall x \in \mathbb{Q})$.

The last result indicates a suitable definition of '*irrational powers*', for if $x \in \mathbb{R} - \mathbb{Q}$ we define e^x to be $\exp(x)$. Since we then have $\exp(x) = e^x$ for all $x \in \mathbb{R}$, we can now discard the $\exp(x)$ notation in favour of e^x.

In general, if $a > 0$ and $b \in \mathbb{R} - \mathbb{Q}$, we define a^b to be $\exp(b \log a)$.

Since the relation $a^b = \exp(b \log a)$ is a theorem for $b \in \mathbb{Q}$, it will now hold for all $b \in \mathbb{R}$, $(a > 0)$.

Other key properties of the exponential function which can now be *proved* are:

(i) $e^x = \lim_{n\to\infty}\left(1+\frac{x}{n}\right)^n \quad (\forall x \in \mathbb{R});$

(ii) $e^x = 1+\frac{1}{1!}x+\frac{1}{2!}x^2+\ldots+\frac{1}{n!}x^n+\ldots, \ \forall x \in \mathbb{R};$

(iii) $\frac{d}{dx}(e^x) = e^x.$

An alternative approach to the logarithmic and exponential functions is based upon (ii) above. We take this as the *definition* of the **exponential function** and derive from it all the other properties quoted.

We define

$$\exp(x) = 1+\frac{1}{1!}x+\frac{1}{2!}x^2+\ldots+\frac{1}{n!}x^n+\ldots,$$

and observe that the power series converges (p. 142) for all real values of x.

The particular value $\exp(1)$ is denoted by e.

Various results in the theory of series are then used to show:

(i) $\exp(a)\exp(b) = \exp(a+b) \quad (\forall a, b \in \mathbb{R});$

(ii) $\exp(x)$ is continuous and differentiable for all x;

(iii) $d/dx\,(\exp(x)) = \exp(x);$

(iv) $\exp(x)$ is a strictly increasing function of x and $\exp(x) > 0$;

(v) $\exp(p) = e^p \quad (\forall p \in \mathbb{Q}).$

We then define $e^x = \exp(x)$ for $x \in \mathbb{R}-\mathbb{Q}$ as before.

It follows from (ii), (iv) and the Inverse Function Theorem (p. 205) that exp has an inverse function – which we call the **logarithmic function** and denote by log – which is also strictly increasing and differentiable.

The domain of log will be $\{x \in \mathbb{R} \mid x > 0\}$ and its codomain $\mathbb{R}$.

It can be readily *shown* that

$$\log(x) = \int_1^x \frac{1}{t}\,dt \quad (x > 0),$$

and the properties of the logarithmic function are now obtained as before.

General logarithms

Using the exponential function we were able (p. 162) to define a^b $(= e^{b \log a})$ for $a > 0$, $b \in \mathbb{R}$. We make use of this in the following definition.

Given $a > 0$ and $a \neq 1$ then we say that y is the **logarithm of x to the base a**, denoted by $y = \log_a x$, if $x = a^y$.

Hyperbolic functions

We define the **hyperbolic sine** (sinh) and the **hyperbolic cosine** (cosh) by

$$\sinh x = \tfrac{1}{2}(e^x - e^{-x}) = \frac{x}{1!} + \frac{x^3}{3!} + \dots + \frac{x^{2n-1}}{(2n-1)!} + \dots,$$

$$\cosh x = \tfrac{1}{2}(e^x + e^{-x}) = 1 + \frac{x^2}{2!} + \dots + \frac{x^{2n}}{(2n)!} + \dots.$$

Sinh and cosh both have domain $\mathbb{R}$; the image of sinh is $\mathbb{R}$, that of cosh is $\{x \in \mathbb{R} \mid x \geqslant 1\}$.

By analogy with the circular trigonometric functions (p. 166), we define the **hyperbolic tangent, cotangent, secant** and **cosecant** by

$$\tanh x = \frac{\sinh x}{\cosh x}, \quad \operatorname{sech} x = \frac{1}{\cosh x},$$

$$\coth x = \frac{\cosh x}{\sinh x}, \quad \operatorname{cosech} x = \frac{1}{\sinh x},$$

whenever the definitions are meaningful.

Circular functions

Several definitions of the sine and cosine functions exist and we give one such below.

We define
$$C(x) = 1 - \frac{x^2}{2!} + \frac{x^4}{4!} - \dots,$$

$$S(x) = x - \frac{x^3}{3!} + \frac{x^5}{5!} - \dots,$$

and observe that the two power series converge for all real values of x.

Now $C(0) > 0$, yet the assumption that $C(x) > 0$ for all $x > 0$ can be shown to lead to a contradiction. Hence, since the set of zeros of a continuous function can be shown to be closed, there is a least positive number x_0 for which $C(x_0) = 0$. We *define* the number π to be $2x_0$.

It is now necessary to show that C and S as defined above possess those properties that characterise the functions which one first encounters in a geometrical setting.

We define the function $\text{cis}: \mathbb{R} \to \mathbb{C}$ by

$$\text{cis}\, x = C(x) + iS(x).$$

Given $z = x+iy \in \mathbb{C}$, the **complex exponential** e^z is defined to be the complex number $e^x \,\text{cis}\, y$.

It can then be shown that, for example,

(i) $e^z = 1 + \frac{1}{1!} z + \frac{1}{2!} z^2 + \frac{1}{3!} z^3 + \ldots, \quad \forall z \in \mathbb{C},$

(ii) $e^{z+w} = e^z e^w, \quad \forall z, w \in \mathbb{C},$

(iii) $e^z \neq 0 \quad (\forall z \in \mathbb{C}),$

(iv) $e^{z+2\pi i} = e^z \quad (\forall z \in \mathbb{C}),$

(v) $C(x) = \dfrac{e^{ix} + e^{-ix}}{2}, \quad S(x) = \dfrac{e^{ix} - e^{-ix}}{2i},$

(vi) C and S are periodic with period 2π (p. 120),

(vii) $\dfrac{d}{dx}(C(x)) = -S(x), \quad \dfrac{d}{dx}(S(x)) = C(x) \quad (C' = -S, S' = C),$

(viii) $C^2(x) + S^2(x) = 1 \quad (\forall x \in \mathbb{R}),$

(ix) C and S are the functions **cosine** and **sine** with which one is familiar from elementary trigonometry.

If one approaches geometry through algebra, then the above definitions can be made the basis of a definition of the measure of an angle. Briefly, we define a *Euclidean plane* (or *space*) to be a two- (or three-) dimensional vector space over $\mathbb{R}$ (p. 38) having an inner product (p. 90) and an associated length or norm (p. 90). Given two vectors x and y, we define the angle between them as follows. We associate with x and y the number

$$\cos\theta = \frac{x \cdot y}{\|x\| \, \|y\|} \quad (-1 \leqslant \cos\theta \leqslant 1)$$

and define the *angle between x and y* to be the number θ ($0 \leqslant \theta \leqslant \pi$) such that

$$1-\frac{\theta^2}{2!}+\frac{\theta^4}{4!}-\ldots = \frac{x.y}{\|x\|\,\|y\|}.$$

Having defined cosine and sine so that they possess the required properties, we define the remaining circular functions – **tangent**, **cotangent**, **secant** and **cosecant** by

$$\tan x = \frac{\sin x}{\cos x},$$

$$\cot x = \frac{\cos x}{\sin x},$$

$$\sec x = \frac{1}{\cos x},$$

$$\operatorname{cosec} x = \frac{1}{\sin x},$$

whenever these definitions are meaningful.

Complex variable

e^{x+iy} was defined on p. 165 as $e^x \operatorname{cis} y$ and it was noted that this was equivalent to the definition

$$e^z = 1+\frac{1}{1!}z+\frac{1}{2!}z^2+\ldots+\frac{1}{n!}z^n+\ldots.$$

Note. (i) $e^{i\theta} = \cos\theta + i\sin\theta$, $\theta \in \mathbb{R}$, $|e^{i\theta}| = 1$ and hence, if $r = |z|$ and $\theta = \arg z$ (pp. 68–9), then $z = re^{i\theta}$.

(ii) As a particular case of the general result stated on p. 73, there are exactly n distinct numbers $z_1, \ldots, z_n$ which satisfy the equation $z^n = 1$. These numbers are known as the nth **roots of unity** and are given by $z_k = e^{i\phi_k}$ where $\phi_k = 2\pi k/n$ ($k = 1, 2, \ldots, n$).

It was noted on p. 165 that $e^z \neq 0$ for any $z \in \mathbb{C}$. However, if $z \in \mathbb{C} - \{0\}$ then there always exists $w \in \mathbb{C}$ such that $e^w = z$. This enables us to make the following definition.

Let $z \neq 0$ be a given complex number and let $w \in \mathbb{C}$ be such that $e^w = z$. The number w is then said to be a **logarithm** of z.

w is not defined uniquely, but if w_1 and w_2 are both logarithms of z then it can be shown that $w_1 - w_2 = 2n\pi i$ for some $n \in \mathbb{Z}$.

One particular logarithm of z is given by

$$w = \log|z| + i\arg(z)$$

and this value is called the **principal logarithm** of z and is denoted by $\operatorname{Log} z$.

Note. (i) The notation given above is by no means universally followed. Some authors denote the *set* of logarithms of z by $\operatorname{Log} z$ (in which case 'Log' is what is traditionally termed a '*many-valued function*') and denote the principal value of the logarithm by $\log z$.

(ii) It will no longer be true that

$$\operatorname{Log}(z_1 z_2) \quad \text{is equal to} \quad \operatorname{Log} z_1 + \operatorname{Log} z_2.$$

The two terms will, however, be congruent modulo $2\pi i$.

Complex circular functions

Given a complex number z we define

$$\cos z = \frac{e^{iz} + e^{-iz}}{2}, \quad \sin z = \frac{e^{iz} - e^{-iz}}{2i}.$$

If $z = x + iy$, we then have

$$\cos z = \cos x \cosh y - i \sin x \sinh y,$$

$$\sin z = \sin x \cosh y + i \cos x \sinh y.$$

See also: De Moivre's Theorem (p. 202); Euler's formula (p. 203).

34 Vector algebra

Vector spaces and, in particular, inner-product spaces (p. 90) were considered, from an algebraic standpoint, earlier in this book. In this section we consider a vector space of great physical importance – the space of *Euclidean vectors* or *geometrical vectors*.

A line segment AB in $\mathbb{R}^3$ is said to be **directed** when one of its end-points is designated an **initial point** and the other a **terminal point**. We indicate that AB is a directed line segment with initial point A and terminal point B by writing $\overrightarrow{AB}$.

If we denote the set of all directed line segments (including degenerate segments such as $\overrightarrow{AA}$) in three-dimensional Euclidean space by S, then we can define an equivalence relation, E, on S by:

The directed line segments $\overrightarrow{AB}$ and $\overrightarrow{CD}$ are said to be equivalent modulo E if, relative to some rectangular Cartesian coordinate system with origin O, the coordinates of their end-points satisfy

$$b_i - a_i = d_i - c_i \quad (i = 1, 2, 3).$$

The quotient set S/E, which we shall denote by $\mathbb{V}$, is the set of **Euclidean vectors.**

A **vector** then is an equivalence class of directed line segments; two line segments being equivalent if they are of equal length and are parallel in the same sense.

Included in every equivalence class $\mathbf{v}$ of line segments will be exactly one segment, $\overrightarrow{OT}$ say, which has its initial point at the origin O.

If we now associate the vector $\mathbf{v}$ with the point T, we obtain a mapping from $\mathbb{V}$ to $\mathbb{R}^3$ which can be shown to be a bijection.

The inverse of this bijection, ρ, say, which is a map from $\mathbb{R}^3$ to $\mathbb{V}$ can then be used to induce the vector space structure from $\mathbb{R}^3$ to $\mathbb{V}$, e.g. addition in $\mathbb{V}$ is defined by

$$\mathbf{v}_1 + \mathbf{v}_2 = \rho(\rho^{-1}(\mathbf{v}_1) + \rho^{-1}(\mathbf{v}_2)).$$

The algebraic operations thus induced have simple geometrical interpretations.

Multiplying $\mathbf{x}$ by a positive scalar λ we obtain the class of directed segments which have length λ times that of members of $\mathbf{x}$ and which are parallel to members of $\mathbf{x}$. Multiplying $\mathbf{x}$ by -1 we obtain the class of segments which have the same length but have opposite direction.

The *sum* $\mathbf{x}+\mathbf{y}$ of two vectors is formed in accordance with the *parallelogram law*, i.e. if $\overrightarrow{OA} \in \mathbf{x}$ and $\overrightarrow{OB} \in \mathbf{y}$ then $\mathbf{x}+\mathbf{y}$ is the vector containing OC where $OACB$ is a parallelogram.

The geometric interpretation of the *scalar product* (i.e., the inner product (see note on p. 165)) is that

$$\mathbf{x}.\mathbf{y} = |\mathbf{x}|\,|\mathbf{y}|\cos\theta,$$

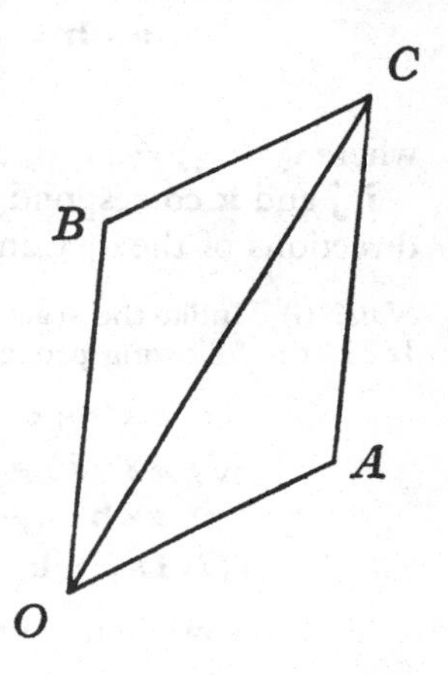

where $|\mathbf{x}|$ denotes the 'length' of the vector $\mathbf{x}$ (p. 90) (which will be the length of all the directed line segments in $\mathbf{x}$) and θ is the angle which the directed line segments in $\mathbf{x}$ make with those in $\mathbf{y}$.

Note. If T has coordinates (t_1, t_2, t_3) then the direction of $\overrightarrow{OT}$ (and of the vector $\mathbf{t}$ to which it belongs) is described by its **direction cosines**, namely,

$$\frac{t_1}{|\mathbf{t}|},\quad \frac{t_2}{|\mathbf{t}|}\quad \text{and}\quad \frac{t_3}{|\mathbf{t}|}.$$

The direction cosines of the zero vector $\mathbf{O}$ corresponding to the segment $\overrightarrow{OO}$ are not defined.

The approach given above is not the historical one and many authors prefer to base definitions of the operations of addition and multiplication by a scalar upon physical considerations and then to show that the resulting structure satisfies the axioms for a vector space given on p. 38.

More confusing differences can arise when the word 'vector' is considered to be interchangeable with 'directed line segment'. In practice, however, one does not always wish to differentiate between an equivalence class and its members (compare the way in which 'rational number' and 'fraction' are often considered interchangeable, whereas the former is an equivalence class of fractions).

Given two vectors $\mathbf{a} = (a_1, a_2, a_3)$, $\mathbf{b} = (b_1, b_2, b_3) \in \mathbb{R}^3$ we define their **vector product (cross product** or **outer product)** denoted by $\mathbf{a}\times\mathbf{b}$ (or $\mathbf{a}\wedge\mathbf{b}$) to be the vector

$$(a_2b_3 - b_3a_2,\ a_3b_1 - a_1b_3,\ a_1b_2 - a_2b_1).$$

Since $\mathbb{V}$ and $\mathbb{R}^3$ are isomorphic vector spaces all the definitions we give for vectors in $\mathbb{R}^3$ apply equally to those in $\mathbb{V}$.

This definition is more easily remembered if we extend the concept of a determinant (p. 52) by permitting formal expansions of determinants containing both scalar and vector elements, and write

$$\mathbf{a}\times\mathbf{b} = \begin{vmatrix} \mathbf{i} & \mathbf{j} & \mathbf{k} \\ a_1 & a_2 & a_3 \\ b_1 & b_2 & b_3 \end{vmatrix} = \begin{vmatrix} a_1 & a_2 & a_3 \\ b_1 & b_2 & b_3 \\ \mathbf{i} & \mathbf{j} & \mathbf{k} \end{vmatrix}$$

where $\mathbf{i}(=\mathbf{e}_1) = (1, 0, 0)$, $\mathbf{j}(=\mathbf{e}_2) = (0, 1, 0)$ and $\mathbf{k}(=\mathbf{e}_3) = (0, 0, 1)$.

i, **j** and **k** correspond geometrically to vectors of unit length in the directions of the x, y and z axes respectively.

Note. (i) Unlike the scalar product, the vector product is not commutative. It has the following properties (which characterise it uniquely):

(*a*) $\mathbf{a}\times(\mathbf{b}+\mathbf{c}) = \mathbf{a}\times\mathbf{b}+\mathbf{a}\times\mathbf{c}$ (*distributivity*);

(*b*) $\mathbf{a}\times(\lambda\mathbf{b}) = \lambda(\mathbf{a}\times\mathbf{b})$ (all $\lambda\in\mathbb{R}$);

(*c*) $\mathbf{a}\times\mathbf{b} = -\mathbf{b}\times\mathbf{a}$ (*anticommutativity*);

(*d*) $\mathbf{i}\times\mathbf{j} = \mathbf{k}$, $\mathbf{j}\times\mathbf{k} = \mathbf{i}$, $\mathbf{k}\times\mathbf{i} = \mathbf{j}$ (cf. p. 70).

(ii) It follows from (*c*) that $\mathbf{a}\times\mathbf{a} = \mathbf{o}$ for all **a**.

Geometrically, $\mathbf{a}\times\mathbf{b}$ is the vector $|\mathbf{a}|\,|\mathbf{b}|\sin\theta\,\mathbf{c}$, where θ is the angle between the directions of **a** and **b**, and **c** is a vector of unit length perpendicular to both **a** and **b** and such that **a**, **b**, **c** form a *right-handed system* of vectors. (The definition of a **right-handed system** is usually given in terms of physical concepts which have no mathematical equivalents. A possible solution is to say that the ordered triple **a**, **b**, **c** form a right-handed system if and only if

$$\begin{vmatrix} a_1 & a_2 & a_3 \\ b_1 & b_2 & b_3 \\ c_1 & c_2 & c_3 \end{vmatrix} > 0.$$

In particular, **i**, **j** and **k** will form a right-handed system and we can then deduce the physical properties of a general right-handed system from those physical properties with which we endow the vectors **i**, **j** and **k** (see figure).) Geometrically, the expression $|\mathbf{a}|\,|\mathbf{b}|\sin\theta$ represents the area of the parallelogram determined by **a** and **b**.

Example. Let $\mathbf{a} = 2\mathbf{i}+3\mathbf{j}-\mathbf{k}$ and $\mathbf{b} = \mathbf{i}-2\mathbf{j}+\mathbf{k}$, i.e. **a** and **b** are the vectors (relative to the basis **i**, **j**, **k**) $(2, 3, -1)$ and $(1, -2, 1)$. Then the *sum*, $\mathbf{a}+\mathbf{b}$, is $3\mathbf{i}+\mathbf{j}$, $2\mathbf{a}$ is the vector $4\mathbf{i}+6\mathbf{j}-2\mathbf{k}$, the *scalar product* $\mathbf{a}.\mathbf{b}$ is the number

$2.1+3.(-2)+(-1).1 = -5$, the *length* of **a** is $\sqrt{14}$, the *direction cosines* of **a** are $2/\sqrt{14}$, $3/\sqrt{14}$ and $-1/\sqrt{14}$, and the *vector product* $\mathbf{a}\times\mathbf{b}$ is the vector

$$\mathbf{c} = \begin{vmatrix} \mathbf{i} & \mathbf{j} & \mathbf{k} \\ 2 & 3 & -1 \\ 1 & -2 & 1 \end{vmatrix} = \mathbf{i}-3\mathbf{j}-7\mathbf{k}.$$

(Note that $\mathbf{a}.\mathbf{c} = \mathbf{b}.\mathbf{c} = 0$.)

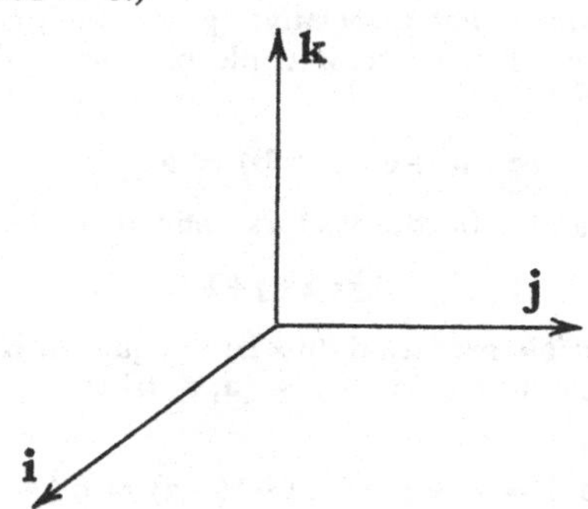

Triple products

Since $\mathbf{a}\times\mathbf{b}$ is a vector it is possible, as was not the case with the scalar product $\mathbf{a}.\mathbf{b}$, to consider vector and scalar products involving $(\mathbf{a}\times\mathbf{b})$.

The scalar product of **a** with the vector $\mathbf{b}\times\mathbf{c}$, i.e. $\mathbf{a}.(\mathbf{b}\times\mathbf{c})$, is known as a **scalar triple product** (or, less happily, as a **triple scalar product**). It will be seen that the scalar triple product $\mathbf{a}.(\mathbf{b}\times\mathbf{c})$ can, without ambiguity, be written $\mathbf{a}.\mathbf{b}\times\mathbf{c}$. It is often denoted by $[\mathbf{a}, \mathbf{b}, \mathbf{c}]$.

Note. (i) It is a simple consequence of the definitions of the scalar and vector products that

$$\mathbf{a}.(\mathbf{b}\times\mathbf{c}) = [\mathbf{a}, \mathbf{b}, \mathbf{c}] = \begin{vmatrix} a_1 & a_2 & a_3 \\ b_1 & b_2 & b_3 \\ c_1 & c_2 & c_3 \end{vmatrix}.$$

(ii) It is easily shown from (i) that (cf. note (i*a*), p. 53)

$$[\mathbf{a}, \mathbf{b}, \mathbf{c}] = [\mathbf{b}, \mathbf{c}, \mathbf{a}] = [\mathbf{c}, \mathbf{a}, \mathbf{b}] = -[\mathbf{a}, \mathbf{c}, \mathbf{b}] = -[\mathbf{b}, \mathbf{a}, \mathbf{c}] = -[\mathbf{c}, \mathbf{b}, \mathbf{a}].$$

Geometrically, the absolute value of $[\mathbf{a}, \mathbf{b}, \mathbf{c}]$ represents the volume of the parallelepiped which has **a**, **b** and **c** for concurrent sides.

$[\mathbf{a}, \mathbf{b}, \mathbf{c}]$ will be positive if the ordered triple **a**, **b**, **c** forms a right-handed system. If **a**, **b**, **c** are coplanar then $[\mathbf{a}, \mathbf{b}, \mathbf{c}] = 0$, i.e. **a**, **b**, **c** are linearly independent (p. 40) in $\mathbb{R}^3$ if and only if $[\mathbf{a}, \mathbf{b}, \mathbf{c}] \neq 0$.

The vector product of $\mathbf{a}$ with the vector $(\mathbf{b}\times\mathbf{c})$ is known as the **vector triple product** (or **triple vector product**).

Note. It is a consequence of the definition that

(*a*) $\mathbf{a}\times(\mathbf{b}\times\mathbf{c}) = (\mathbf{a}.\mathbf{c})\mathbf{b}-(\mathbf{a}.\mathbf{b})\mathbf{c}$
$\neq (\mathbf{a}\times\mathbf{b})\times\mathbf{c},$

i.e. the vector product is *not* associative (p. 17) and, accordingly, brackets must always be inserted in a vector triple product if the product is to be meaningful.

(*b*) $\mathbf{a}\times(\mathbf{b}\times\mathbf{c})+\mathbf{b}\times(\mathbf{c}\times\mathbf{a})+\mathbf{c}\times(\mathbf{a}\times\mathbf{b}) = \mathbf{0}.$

Example. Let $\mathbf{a}$, $\mathbf{b}$ and $\mathbf{c}$ be the vectors defined in the example on p. 170 and let

$$\mathbf{d} = \mathbf{i}+\mathbf{j}+\mathbf{k}.$$

Then the *scalar triple product* $\mathbf{d}.(\mathbf{a}\times\mathbf{b})$ is equal to $\mathbf{d}.\mathbf{c} = -9$.
The *scalar triple product* $\mathbf{a}.(\mathbf{d}\times\mathbf{b}) = [\mathbf{a}, \mathbf{d}, \mathbf{b}]$ is

$$\begin{vmatrix} 2 & 3 & -1 \\ 1 & 1 & 1 \\ 1 & -2 & 1 \end{vmatrix} = 2(+3)-3(0)-1(-3) = 9\ (= -[\mathbf{d}, \mathbf{a}, \mathbf{b}]).$$

The *vector triple product* $\mathbf{d}\times(\mathbf{a}\times\mathbf{b})$ is equal to $\mathbf{d}\times\mathbf{c}$, i.e. to

$$\begin{vmatrix} \mathbf{i} & \mathbf{j} & \mathbf{k} \\ 1 & 1 & 1 \\ 1 & -3 & -7 \end{vmatrix} = -4\mathbf{i}+8\mathbf{j}-4\mathbf{k}.$$

(Note that $(\mathbf{d}.\mathbf{b})\mathbf{a}-(\mathbf{d}.\mathbf{a})\mathbf{b} = 0\mathbf{a}-4\mathbf{b} = -4\mathbf{b} = \mathbf{d}\times(\mathbf{a}\times\mathbf{b})$.

Since $\mathbf{d}.\mathbf{b} = 0$ and $\mathbf{d} \neq \mathbf{0} \neq \mathbf{b}$, it follows that $\mathbf{d}$ and $\mathbf{b}$ are orthogonal.)

35 Vector calculus

In applications of mathematics one is often concerned with the behaviour of **vector fields** (or **field vectors**), i.e. vector-valued functions (p. 126) defined upon some subset of $\mathbb{R}^n$ (for some $n \in \mathbb{N}$).

As before (cf. p. 16 and p. 126), if the codomain of the vector-valued function $\mathbf{f}$ is $\mathbb{R}^m$, then we can write

$$\mathbf{f}(\mathbf{x}) = f_1(\mathbf{x})\,\mathbf{e}_1 + f_2(\mathbf{x})\,\mathbf{e}_2 + \ldots + f_m(\mathbf{x})\,\mathbf{e}_m,$$

where $f_1, f_2, \ldots, f_m$ are real-valued functions and $(\mathbf{e}_1, \ldots, \mathbf{e}_m)$ is some basis of $\mathbb{R}^m$. To distinguish between a vector field and its real-valued components, the latter are often referred to as **scalar fields**.

Let ϕ be a real-valued function defined on an open set S in $\mathbb{R}^3$. We define the **gradient** of ϕ, denoted by $\nabla\phi$ or **grad** ϕ, to be the vector-valued function defined by

$$\nabla\phi(\mathbf{a}) = \left.\frac{\partial\phi}{\partial x_1}\right|_{\mathbf{a}} \mathbf{i} + \left.\frac{\partial\phi}{\partial x_2}\right|_{\mathbf{a}} \mathbf{j} + \left.\frac{\partial\phi}{\partial x_3}\right|_{\mathbf{a}} \mathbf{k}$$

at each point $\mathbf{a} \in S$ at which the partial derivatives (p. 126) exist.

More generally, given $\phi : S \subset \mathbb{R}^n \to \mathbb{R}$, we define $\nabla\phi(\mathbf{a})$ to be the vector

$$(D_1\phi|_{\mathbf{a}},\quad D_2\phi|_{\mathbf{a}}, \ldots, D_n\phi|_{\mathbf{a}}) \quad \text{(p. 126)},$$

i.e. $\nabla\phi(\mathbf{a})$ is the Jacobian matrix (p. 127) of ϕ at $\mathbf{a}$ (or its transpose according to the convention adopted).

A geometrical interpretation of the gradient is a consequence of the following. The equation $\phi(\mathbf{x}) = c$ often represents a **surface** (p. 179) S_c in $\mathbb{R}^3$ and given any $\mathbf{a} \in S$ there will be some $c \in \mathbb{R}$ such that the surface S_c passes through $\mathbf{a}$. If $\nabla\phi(\mathbf{a}) \neq \mathbf{o}$, then S_c has a *tangent plane* at $\mathbf{a}$ to which the vector $\nabla\phi(\mathbf{a})$ is *normal.*

Moreover, if $\mathrm{d}_{\mathbf{a}}\phi$ denotes the differential of ϕ at $\mathbf{a}$ (p. 129) then we see that

$$\mathrm{d}_{\mathbf{a}}\phi(\mathbf{x}) = \nabla\phi(\mathbf{a}).\mathbf{x},$$

from which it follows that the direction of the gradient at a point $\mathbf{a}$ is the direction in which ϕ is increasing most rapidly, and that the magnitude of the gradient gives the rate of increase.

The scalar field ϕ whose gradient is $\nabla\phi$ is called the **potential function** of the vector field $\nabla\phi$. The corresponding surfaces $\phi(\mathbf{x}) = c$, $c \in \mathbb{R}$, are called **equipotential surfaces.**

Note. (i) It is often convenient to think of the vector operator ∇ (**nabla**) as though it were a symbolic vector, namely,

$$\left(\frac{\partial}{\partial x_1}, \frac{\partial}{\partial x_2}, \frac{\partial}{\partial x_3}\right) \quad \text{or} \quad (D_1, D_2, D_3).$$

(ii) It is easily shown that

$$(a)\ \nabla(\phi+\theta) = \nabla\phi+\nabla\theta, \quad (b)\ \nabla(\theta\phi) = \theta\nabla\phi+\phi\nabla\theta.$$

(iii) $\nabla\phi(\mathbf{x})$ represents a definite vector which is independent of the choice of our rectangular coordinate system (orthonormal basis, p. 90).

Example. Consider $\phi : (x, y, z) \mapsto \dfrac{1}{\sqrt{(x^2+y^2+z^2)}} = \dfrac{1}{r} \quad ((x, y, z) \neq (0, 0, 0)).$

Then

$$\nabla\phi(x_0, y_0, z_0) = -\frac{x_0}{(x_0^2+y_0^2+z_0^2)^{\frac{3}{2}}}\mathbf{i}-\frac{y_0}{(x_0^2+y_0^2+z_0^2)^{\frac{3}{2}}}\mathbf{j}-\frac{z_0}{(x_0^2+y_0^2+z_0^2)^{\frac{3}{2}}}\mathbf{k}$$

$$= -\frac{1}{r_0^3}(x_0\mathbf{i}+y_0\mathbf{j}+z_0\mathbf{k}).$$

Note that $\phi(\mathbf{x}) = c$ represents a sphere with centre the origin and that $\nabla\phi(\mathbf{x}_0)$ is a vector along the radius through (x_0, y_0, z_0), i.e. $\nabla\phi(\mathbf{x}_0)$ is the *normal* at $\mathbf{x}_0$ to the surface $\phi(\mathbf{x}) = c$.

Conversely, given the vector field

$$\mathbf{f} : \mathbf{x} \mapsto -\left(\frac{x_1}{|\mathbf{x}|^3}, \frac{x_2}{|\mathbf{x}|^3}, \frac{x_3}{|\mathbf{x}|^3}\right) \quad (|\mathbf{x}| \neq 0),$$

then the *potential function* of $\mathbf{f}$ is $1/r$, and the *equipotential surfaces* are spheres with centre the origin.

Let $\mathbf{f}$ be a vector-valued function defined on an open set S in $\mathbb{R}^3$, with codomain $\mathbb{R}^3$ and components f_1, f_2, f_3. We denote by **curl f** the vector-valued function from S to $\mathbb{R}^3$ defined by

$$\operatorname{curl}\mathbf{f} = \left(\frac{\partial f_3}{\partial x_2}-\frac{\partial f_2}{\partial x_3}, \frac{\partial f_1}{\partial x_3}-\frac{\partial f_3}{\partial x_1}, \frac{\partial f_2}{\partial x_1}-\frac{\partial f_1}{\partial x_2}\right)$$

$$= (D_2 f_3 - D_3 f_2, D_3 f_1 - D_1 f_3, D_1 f_2 - D_2 f_1),$$

whenever the partial derivatives exist.

Using the 'vector' ∇ (note (i) above), curl $\mathbf{f}$ may be remembered as

$$\operatorname{curl}\mathbf{f} = \nabla\times\mathbf{f} = \begin{vmatrix} \mathbf{i} & \mathbf{j} & \mathbf{k} \\ \dfrac{\partial}{\partial x_1} & \dfrac{\partial}{\partial x_2} & \dfrac{\partial}{\partial x_3} \\ f_1 & f_2 & f_3 \end{vmatrix},$$

where, as on p. 170, the determinant is to be evaluated formally.

In physical applications, curl represents some measure of rotation (older texts often describe curl **f** as the **rotation** of **f** and denote it by **rotf**) and for this reason a vector field **f** satisfying curl**f** = **o**, i.e. curl**f**(**x**) = **o** for all $\mathbf{x} \in S$, is said to be **irrotational**.

Note. (i) It can be shown that

$$(a)\ \operatorname{curl}(\mathbf{f}+\mathbf{g}) = \operatorname{curl}\mathbf{f}+\operatorname{curl}\mathbf{g},$$
$$(b)\ \operatorname{curl}(\phi\mathbf{f}) = \phi\operatorname{curl}\mathbf{f}+\nabla\phi\times\mathbf{f},$$
$$(c)\ \operatorname{curl}(\operatorname{grad}\phi) = \mathbf{o}.$$

(For brevity, in this note and the corresponding one for *divergence* on p. 176, the exact conditions under which the relations hold have been omitted.)

(ii) curl**f** is independent of the choice of orthonormal basis.

Example. Consider $\mathbf{f}: (x, y, z) \mapsto \left(\frac{x}{r^3}, \frac{y}{r^3}, \frac{z}{r^3}\right) \quad ((x, y, z) \neq (0, 0, 0)).$

Then

$$\operatorname{curl}\mathbf{f}\,(x, y, z) = \nabla\times\mathbf{f}\,(x, y, z) = \begin{vmatrix} \mathbf{i} & \mathbf{j} & \mathbf{k} \\ \frac{\partial}{\partial x} & \frac{\partial}{\partial y} & \frac{\partial}{\partial z} \\ \frac{x}{r^3} & \frac{y}{r^3} & \frac{z}{r^3} \end{vmatrix}$$

$$= \mathbf{i}\left(\frac{\partial}{\partial y}\left(\frac{z}{r^3}\right) - \frac{\partial}{\partial z}\left(\frac{y}{r^3}\right)\right) + \dots$$

$$= \mathbf{i}\left(\frac{3zy}{r^5} - \frac{3yz}{r^5}\right) + \dots$$

$$= \mathbf{o}$$

(cf. the example on p. 174 and note (i) (*c*) above). See also example (ii) on p. 176.

Let $\mathbf{f} = (f_1, f_2, f_3)$ be a vector-valued function defined on S an open set in $\mathbb{R}^3$. The **divergence** of **f**, denoted by **div f**, is defined to be the scalar field given by

$$\operatorname{div}\mathbf{f} = \frac{\partial f_1}{\partial x_1} + \frac{\partial f_2}{\partial x_2} + \frac{\partial f_3}{\partial x_3} = D_1 f_1 + D_2 f_2 + D_3 f_3$$

whenever the partial derivatives exist.

Using the nabla notation (p. 174) we have

$$\operatorname{div}\mathbf{f} = \nabla.\mathbf{f}.$$

Like that of the gradient, this definition can be extended to functions defined on $\mathbb{R}^n$ where $n > 3$.

In physical applications the divergence of the vector field **f** is a measure of the rate at which fluid, say, is flowing away from the immediate vicinity of a point. A vector field satisfying div**f** = 0, i.e. div**f**(**x**) = 0 for all $\mathbf{x} \in S$, is said to be **solenoidal**.

Note. (i) It can be shown that

$$(a)\ \operatorname{div}(\mathbf{f}+\mathbf{g}) = \operatorname{div}\mathbf{f}+\operatorname{div}\mathbf{g},$$
$$(b)\ \operatorname{div}(\phi\mathbf{f}) = \phi\operatorname{div}\mathbf{f}+\nabla\phi.\mathbf{f},$$
$$(c)\ \operatorname{div}(\operatorname{curl}\mathbf{f}) = 0.$$

The exact conditions under which these conditions hold have been omitted for the sake of brevity.

(ii) $\operatorname{div}\mathbf{f}$ is independent of the choice of orthonormal basis.

(iii) The *divergence of a gradient* is of particular physical importance. It is expressed symbolically by $\nabla.\nabla\phi$ or, more briefly, by $\nabla^2\phi$ (del squared ϕ) or $\Delta\phi$. ∇^2 is called the **Laplacian operator** and the partial differential equation $\nabla^2\phi = 0$ is known as **Laplace's equation.** If a function ϕ satisfies Laplace's equation on an open set S, it is said to be **harmonic** on that set. The domain of the Laplacian operator can be extended to vector fields by defining $\nabla^2\mathbf{f}$, where $\mathbf{f} = (f_1, f_2, f_3)$ to be $(\nabla^2 f_1, \nabla^2 f_2, \nabla^2 f_3)$. It can then be shown that

$$\operatorname{curl}(\operatorname{curl}\mathbf{f}) = \operatorname{grad}(\operatorname{div}\mathbf{f}) - \nabla^2\mathbf{f}.$$

(Care must be taken here to differentiate between the operators $\nabla\times(\nabla\times\quad)$, $\nabla(\nabla.\quad)$ and ∇^2.)

Examples. (i) Consider $\mathbf{f}: (x, y, z) \mapsto \left(\frac{x}{r^3}, \frac{y}{r^3}, \frac{z}{r^3}\right) \quad ((x, y, z) \neq (0, 0, 0)).$

Then

$$\begin{aligned}\operatorname{div}\mathbf{f}(x, y, z) = \nabla.\mathbf{f}(x, y, z) &= \frac{\partial}{\partial x}\left(\frac{x}{r^3}\right)+\frac{\partial}{\partial y}\left(\frac{y}{r^3}\right)+\frac{\partial}{\partial z}\left(\frac{z}{r^3}\right)\\ &= \frac{1}{r^3}-\frac{3x^2}{r^5}+\frac{1}{r^3}-\frac{3y^2}{r^5}+\frac{1}{r^3}-\frac{3z^2}{r^5}\\ &= \frac{3}{r^3}-\frac{3}{r^5}(x^2+y^2+z^2)\\ &= 0.\end{aligned}$$

Hence, from the example on p. 174, $\phi: (x, y, z) \mapsto 1/r$ satisfies $\nabla^2\phi = 0$ and is a *harmonic function* on any open set of $\mathbb{R}^3$ excluding the origin.

(ii) Consider $\mathbf{f}: (x, y, z) \mapsto (x^2y, y^2-xyz, xz^2)$.

Then

$$\operatorname{curl}\mathbf{f}(\mathbf{x}) = \nabla\times\mathbf{f}(\mathbf{x}) = \begin{vmatrix}\mathbf{i} & \mathbf{j} & \mathbf{k}\\ \frac{\partial}{\partial x} & \frac{\partial}{\partial y} & \frac{\partial}{\partial z}\\ x^2y & y^2-xyz & xz^2\end{vmatrix}$$

$$= xy\mathbf{i}-z^2\mathbf{j}-(yz+x^2)\mathbf{k}$$

and

$$\operatorname{div}(\operatorname{curl}\mathbf{f}(\mathbf{x})) = \nabla.(\nabla\times\mathbf{f}(\mathbf{x})) = y-y = 0 \quad \text{(cf. note (i)(c) above).}$$

See also: Mean Value Theorems (p. 208).

36 Line and surface integrals

Line integrals

Let $\boldsymbol{\alpha}$ be a curve in $\mathbb{R}^n$ defined on the closed interval $\leqslant a, b \geqslant$ (p. 150), $P = \{a_0, \ldots, a_m\}$ be a partition of $\leqslant a, b \geqslant$ and $\mathbf{f}$ be a function mapping the image of $\boldsymbol{\alpha}$ (in $\mathbb{R}^n$) to $\mathbb{R}^n$. (See note (ii) below.)

We denote by $S(P, \mathbf{f}, \boldsymbol{\alpha})$ the sum of the inner products (p. 90)

$$\sum_{i=1}^{m} (\mathbf{f}(\boldsymbol{\alpha}(v_i)) . (\boldsymbol{\alpha}(a_i) - \boldsymbol{\alpha}(a_{i-1})),$$

where v_i is some chosen point in the interval $\leqslant a_{i-1}, a_i \geqslant$ ($i = 1, \ldots, m$).

The **line integral** (or **curvilinear integral**) of $\mathbf{f}$ along the path $\boldsymbol{\alpha}$ exists if there is a real number A for which, given any $\epsilon > 0$, there is a partition P_ϵ of $\leqslant a, b \geqslant$ such that for every refinement P of P_ϵ and for every choice of the points v_i, we have

$$|A - S(P, \mathbf{f}, \boldsymbol{\alpha})| < \epsilon.$$

If such a number exists, it is uniquely determined and is denoted by the symbol

$$\int_{\boldsymbol{\alpha}} \mathbf{f} . \mathrm{d}\boldsymbol{\alpha},$$

or, writing $\mathbf{f} = (f_1, \ldots, f_n)$ (p. 126), by

$$\int_{\boldsymbol{\alpha}} (f_1 \mathrm{d}\alpha_1 + f_2 \mathrm{d}\alpha_2 + \ldots + f_n \mathrm{d}\alpha_n).$$

In the case $n = 3$, $\boldsymbol{\alpha}$ is often described in terms of a parameter t by

$$\boldsymbol{\alpha}(t) = (x(t), y(t), z(t)), \quad a \leqslant t \leqslant b.$$

The sum $S(P, \mathbf{f}, \boldsymbol{\alpha})$ will then have the form

$$\sum_{i=1}^{m} [f_1(\boldsymbol{\alpha}(v_i))(x(t_i) - x(t_{i-1})) + f_2(\boldsymbol{\alpha}(v_i))(y(t_i) - y(t_{i-1})) + f_3(\boldsymbol{\alpha}(v_i))(z(t_i) - z(t_{i-1}))]$$

and the line integral will be denoted by

$$\int_{\boldsymbol{\alpha}} (f_1 \mathrm{d}x + f_2 \mathrm{d}y + f_3 \mathrm{d}z).$$

Note. (i) The line integral will always exist if $\boldsymbol{\alpha}$ is rectifiable (p. 151) and $\mathbf{f}$ is continuous on (the image of) $\boldsymbol{\alpha}$.

(ii) The line integral is dependent upon the direction in which the curve is traversed (which will affect the sign of the integral).

(iii) It will be observed that in the description given above for the case $n = 3$ we are effectively summing three Riemann–Stieltjes integrals, namely that of f_1 with respect to x, f_2 with respect to y and f_3 with respect to z. In general, if $\mathbf{f}$ is bounded on $\boldsymbol{\alpha}$, then

$$\int_{\boldsymbol{\alpha}} \mathbf{f}.\mathrm{d}\boldsymbol{\alpha} = \sum_{i=1}^{n} \int_a^b f_i(\boldsymbol{\alpha}(t))\, \mathrm{d}\alpha_i(t).$$

Again, using the relation between the Riemann integral and the Riemann–Stieltjes integral, the right-hand side can be written in terms of Riemann integrals provided $\boldsymbol{\alpha}$ is piecewise smooth (p. 151) to give

$$\int_{\boldsymbol{\alpha}} \mathbf{f}.\mathrm{d}\boldsymbol{\alpha} = \sum_{i=1}^{n} \int_a^b f_i(\boldsymbol{\alpha}(t))\, \alpha_i'(t)\, \mathrm{d}t.$$

In particular, when $n = 3$ we have

$$\int_{\boldsymbol{\alpha}} \mathbf{f}.\mathrm{d}\boldsymbol{\alpha} = \int_a^b \left(f_1 \frac{\mathrm{d}x}{\mathrm{d}t} + f_2 \frac{\mathrm{d}y}{\mathrm{d}t} + f_3 \frac{\mathrm{d}z}{\mathrm{d}t}\right) \mathrm{d}t.$$

(iv) If $\boldsymbol{\alpha}$ is a positively oriented, rectifiable Jordan curve in $\mathbb{R}^2$ (p. 158 and p. 151), then the line integral along $\boldsymbol{\alpha}$ is denoted by

$$\oint_{\boldsymbol{\alpha}}.$$

(v) The connection between contour integrals (p. 157) and line integrals is illustrated by the fact that if

$$f(z) = f_1(x, y) + \mathrm{i} f_2(x, y)$$

where f_1 and f_2 are real valued functions defined on $\boldsymbol{\alpha}$, a curve in $\mathbb{R}^2$ given by

$$\boldsymbol{\alpha}(t) = x(t) + \mathrm{i}y(t), \quad a \leqslant t \leqslant b,$$

and if the contour integral of f along $\boldsymbol{\alpha}$ exists, then

$$\int_{\boldsymbol{\alpha}} f(z)\,\mathrm{d}z = \int_{\boldsymbol{\alpha}} (f_1\,\mathrm{d}x - f_2\,\mathrm{d}y) + \mathrm{i} \int_{\boldsymbol{\alpha}} (f_2\,\mathrm{d}x + f_1\,\mathrm{d}y).$$

Example. Consider

$$\mathbf{f}: (x, y) \mapsto \left(\frac{x}{x^2+y^2}, \frac{-y}{x^2+y^2}\right) \quad \text{and} \quad \boldsymbol{\alpha}: t \mapsto (\cos t, \sin t)$$

defined on $\leqslant 0, 2\pi \geqslant$. Then,

$$\oint_\alpha \mathbf{f}.\mathrm{d}\alpha = \int_\alpha (f_1\,\mathrm{d}x + f_2\,\mathrm{d}y)$$

$$= \int_0^{2\pi} \left(\cos t\,\frac{\mathrm{d}x}{\mathrm{d}t} - \sin t\,\frac{\mathrm{d}y}{\mathrm{d}t}\right)\mathrm{d}t$$

$$= \int_0^{2\pi} (-\cos t\,\sin t - \sin t\,\cos t)\,\mathrm{d}t = \int_0^{2\pi} -\sin 2t\,\mathrm{d}t$$

$$= \left[\frac{\cos 2t}{2}\right]_0^{2\pi} = 0.$$

We see that, in particular,

$$\int_\alpha f_1\,\mathrm{d}x = 0, \quad \int_\alpha -f_2\,\mathrm{d}y = 0.$$

However, $\displaystyle\int_\alpha f_2\,\mathrm{d}x = \int_0^{2\pi} (-\sin t)(-\sin t)\,\mathrm{d}t = \frac{1}{2}\int_0^{2\pi} (1-\cos 2t)\,\mathrm{d}t$

$$= \frac{1}{2}\left[t - \frac{\sin 2t}{2}\right]_0^{2\pi} = \pi,$$

and $\displaystyle\int_\alpha f_1\,\mathrm{d}y = \int_0^{2\pi} (\cos t)(\cos t)\,\mathrm{d}t = \pi,$

and so, setting $\displaystyle f(z) = \frac{x}{x^2+y^2} - \frac{\mathrm{i}y}{x^2+y^2} = \frac{1}{z},$

we have $\displaystyle\int_\alpha \frac{1}{z}\,\mathrm{d}z = 2\pi\mathrm{i}$

(see note (v) and compare example, p. 158).

Surface integrals

Let R be a **plane region** in $\mathbb{R}^2$, i.e. the union of a rectifiable Jordan curve, $\mathbf{\Gamma}$ (pp. 150–1), with its interior, and $\mathbf{x}(\mathbf{t}) = ((x_1(\mathbf{t})), x_2(\mathbf{t}), x_3(\mathbf{t}))$ be a vector-valued function which is defined and has continuous partial derivatives for all $\mathbf{t} \in R^o$, an open set in $\mathbb{R}^2$ containing R. The image, S, of R under $\mathbf{x}$ is said to be a **parametric surface** described by $\mathbf{x}$. If $\mathbf{x}$ is an injection (p. 14) on R, then S is said to be a **simple parametric surface** and the image of $\mathbf{\Gamma}$, which will be a rectifiable Jordan curve, is called the **edge** of S.

Note. As in the case of the definition of a *curve* (p. 150), there are advantages to be gained from defining the surface to be the *function* $\mathbf{x}$ rather than its *image*. However, in this instance the traditional nomenclature has been followed.

Example. Consider $\mathbf{x} : \mathbb{R}^2 \to \mathbb{R}^3$ defined by

$$\mathbf{x}(t_1, t_2) = \sin t_1 \cos t_2 \mathbf{i} + \sin t_1 \sin t_2 \mathbf{j} + \cos t_1 \mathbf{k},$$

and let $\quad R \subset \mathbb{R}^2 \;$ be $\; \{(t_1, t_2) | 0 \leqslant t_1 \leqslant \pi/2, 0 \leqslant t_2 \leqslant 2\pi\}$.

R is a *plane region* (see diagram) in $\mathbb{R}^2$ and $\mathbf{\Gamma}$ its boundary is clearly rectifiable being the union of four straight line segments ($\mathbf{\Gamma}_1, \ldots, \mathbf{\Gamma}_4$). $\mathbf{x}$ has continuous partial derivatives in an open set of $\mathbb{R}^2$ containing R and hence S, the image of R under $\mathbf{x}$, is a *parametric surface* in $\mathbb{R}^3$.

We note that $\mathbf{x}(t_1, 0) = \mathbf{x}(t_1, 2\pi)$ and so $\mathbf{x}$ is *not* injective and accordingly S is *not* a simple parametric surface.

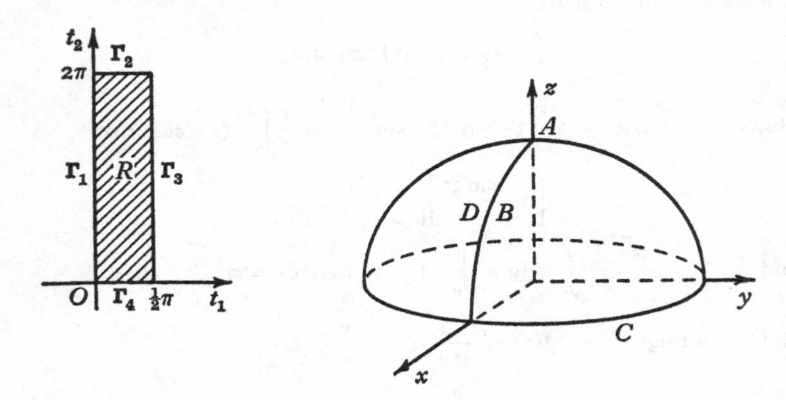

The image of $\mathbf{\Gamma}$ under $\mathbf{x}$ is $A \cup B \cup C \cup D$ where

$A = \{(x, y, z) | x = 0, y = 0, z = 1\}$ (image of $\mathbf{\Gamma}_1$),

$B = \{(x, y, z) | x^2 + z^2 = 1, x \geqslant 0, z \geqslant 0, y = 0\}$ (image of $\mathbf{\Gamma}_2$),

$C = \{(x, y, z) | x^2 + y^2 = 1, z = 0\}$ (image of $\mathbf{\Gamma}_3$),

$D = \{(x, y, z) | x^2 + z^2 = 1, x \geqslant 0, z \geqslant 0, y = 0\}$ (image of $\mathbf{\Gamma}_4$).

We note that if $(x, y, z) \in S$, then $x^2 + y^2 + z^2 = 1$ and $z \geqslant 0$; S is, in fact, a hemisphere.

In this case the image of $\mathbf{\Gamma}$ is not the 'edge' of S in the obvious physical sense.

Of particular interest are those surfaces for which a tangent plane can be defined at almost every point and in order to characterise these surfaces one defines vectors $\mathrm{D}_1\mathbf{x}(\mathbf{t})$ and $\mathrm{D}_2\mathbf{x}(\mathbf{t})$ by

$$\mathrm{D}_1\mathbf{x}(\mathbf{t}) = \mathrm{D}_1 x_1(\mathbf{t})\mathbf{i} + \mathrm{D}_1 x_2(\mathbf{t})\mathbf{j} + \mathrm{D}_1 x_3(\mathbf{t})\mathbf{k},$$
$$\mathrm{D}_2\mathbf{x}(\mathbf{t}) = \mathrm{D}_2 x_1(\mathbf{t})\mathbf{i} + \mathrm{D}_2 x_2(\mathbf{t})\mathbf{j} + \mathrm{D}_2 \mathbf{x}_3(\mathbf{t})\mathbf{k},$$

where $\mathbf{t} = (t_1, t_2) \in R$, and $D_i x_j(\mathbf{t})$ $(i = 1, 2, j = 1, 2, 3)$ denote the partial derivatives in $\mathbb{R}^2$ of the functions $x_1(\mathbf{t}), x_2(\mathbf{t}), x_3(\mathbf{t})$. Those points $\mathbf{x}(\mathbf{t})$ of S where the cross product $D_1\mathbf{x}(\mathbf{t}) \times D_2\mathbf{x}(\mathbf{t})$ is non-zero will possess a **tangent plane** (the plane determined by the two non-zero vectors $D_1\mathbf{x}(\mathbf{t})$ and $D_2\mathbf{x}(\mathbf{t})$) and are called **regular** points. Points which are not regular are said to be **singular**. Those surfaces which possess at most a finite number of singular points are of particular interest.

Note. (i) The cross product $D_1\mathbf{x}(\mathbf{t}) \times D_2\mathbf{x}(\mathbf{t})$ can be written in terms of Jacobians (p. 127) as

$$\frac{\partial(x_2, x_3)}{\partial(t_1, t_2)}\mathbf{i} + \frac{\partial(x_3, x_1)}{\partial(t_1, t_2)}\mathbf{j} + \frac{\partial(x_1, x_2)}{\partial(t_1, t_2)}\mathbf{k}$$
$$= |J_1(\mathbf{t})|\,\mathbf{i} + |J_2(\mathbf{t})|\,\mathbf{j} + |J_3(\mathbf{t})|\,\mathbf{k} \quad \text{say.}$$

(ii) It is a consequence of the Implicit Function Theorem (p. 205) that if, say, $|J_3(\mathbf{t}_0)| \neq 0$, then there is a neighbourhood N of $(x_1(\mathbf{t}_0), x_2(\mathbf{t}_0))$ in $\mathbb{R}^2$ and a function $\phi : \mathbb{R}^2 \to \mathbb{R}$ defined on N such that whenever $(x_1, x_2) \in N$ we have $x_3 = \phi(x_1, x_2)$. If this last equation holds for all points (x_1, x_2, x_3) of S, then one can identify the $t_1 t_2$-plane with the $x_1 x_2$-plane and write

$$\mathbf{x}(\mathbf{t}) = t_1\mathbf{i} + t_2\mathbf{j} + \phi(t_1, t_2)\mathbf{k} \quad ((t_1, t_2) \in R).$$

We then say that S has a one-to-one projection onto the $x_1 x_2$- (or $t_1 t_2$-) plane.

Example. Consider $\mathbf{x}$ as defined in the example on p. 180.

Then

$$D_1\mathbf{x}(\mathbf{t}) = \cos t_1 \cos t_2 \mathbf{i} + \cos t_1 \sin t_2 \mathbf{j} - \sin t_1 \mathbf{k},$$
$$D_2\mathbf{x}(\mathbf{t}) = -\sin t_1 \sin t_2 \mathbf{i} + \sin t_1 \cos t_2 \mathbf{j},$$
$$D_1\mathbf{x}(\mathbf{t}) \times D_2\mathbf{x}(\mathbf{t}) = \sin^2 t_1 \cos t_2 \mathbf{i} + \sin^2 t_1 \sin t_2 \mathbf{j} + \sin t_1 \cos t_1 \mathbf{k}.$$

$D_1\mathbf{x}(\mathbf{t}) \times D_2\mathbf{x}(\mathbf{t})$ is, therefore, zero if and only if $\sin t_1 = 0$, i.e., if $t_1 = 0$. Hence, the singular points of S are those points lying in the image of $\mathbf{\Gamma}_1$. However, $\mathrm{Im}(\mathbf{\Gamma}_1) = \{(0, 0, 1)\}$, i.e. the point $(0, 0, 1)$ is the only *singular point* of S. All other points of S are *regular*. (We note, for use in the example on p. 182, that $D_1\mathbf{x}(\mathbf{t}) \times D_2\mathbf{x}(\mathbf{t}) = \sin t_1 \mathbf{x}(\mathbf{t})$.)

If S is a parametric surface defined by a vector-valued function $\mathbf{x}$ defined on a plane region R, then the **area** of S is defined to be the value of the double integral (p. 160)

$$\int_R |D_1\mathbf{x}(\mathbf{t}) \times D_2\mathbf{x}(\mathbf{t})|\, d(t_1, t_2).$$

Compare the integral for finding the *length* of an arc (p. 150).

Example. We consider further $\mathbf{x}$, R and S as defined in the examples above and on p. 180.

The *area* of S is, therefore,

$$\int_R |D_1\mathbf{x}(\mathbf{t}) \times D_2\mathbf{x}(\mathbf{t})|\, d(t_1, t_2)$$
$$= \int_R |\sin t_1 \mathbf{x}(\mathbf{t})|\, d(t_1, t_2) = \int_0^{2\pi}\left[\int_0^{\frac{1}{2}\pi} \sin t_1\, dt_1\right] dt_2 \quad \text{(see p. 160)},$$
$$= 2\pi.$$

When S is a parametric surface of the type described above, then the two unit vectors *normal* to S at a regular point corresponding to a point $\mathbf{t} \in R$, $\mathbf{n}_1(\mathbf{t})$ and $\mathbf{n}_2(\mathbf{t})$, say, are given by

$$\mathbf{n}_1(\mathbf{t}) = \frac{D_1\mathbf{x}(\mathbf{t}) \times D_2\mathbf{x}(\mathbf{t})}{|D_1\mathbf{x}(\mathbf{t}) \times D_2\mathbf{x}(\mathbf{t})|} \quad \text{and} \quad \mathbf{n}_2(\mathbf{t}) = -\mathbf{n}_1(\mathbf{t}).$$

If, now, $\mathbf{f} = (f_1, f_2, f_3)$ is a vector-valued function defined on $S\ (\subset \mathbb{R}^3)$, we define the **surface integral** of $\mathbf{f}.\mathbf{n}$ (where $\mathbf{n}$ is either $\mathbf{n}_1$ or $\mathbf{n}_2$), denoted by

$$\iint_S \mathbf{f}.\mathbf{n}\, ds,$$

to be the double integral

$$\int_R \mathbf{f}(\mathbf{x}(\mathbf{t})).\mathbf{n}(\mathbf{t})|D_1\mathbf{x}(\mathbf{t}) \times D_2\mathbf{x}(\mathbf{t})|\, d(t_1, t_2)$$

whenever that integral exists.

Note. Alternative definitions of the surface integral can be found in the literature and their equivalence is not always obvious. We indicate two other approaches below.

(i) $\mathbf{n}\,ds$ is often replaced by a 'vector' $\mathbf{ds}$ of magnitude ds (an element of the surface S) and with direction normal to the surface. We then speak of the **surface integral of f over** S and denote it by

$$\iint_S \mathbf{f}.\mathbf{ds}.$$

(ii) As a consequence of note (i) on p. 181 we may write

$$\iint_S \mathbf{f}.\mathbf{n}\, ds = \pm \sum_{i=1}^{3} \int_R f_i(\mathbf{x}(\mathbf{t}))\, |J_i(\mathbf{t})|\, d(t_1, t_2)$$
$$= \pm\left\{\iint_S f_1(x_1, x_2, x_3)\, dx_2\, dx_3 + \iint_S f_2(x_1, x_2, x_3)\, dx_3\, dx_1 + \iint_S f_3(x_1, x_2, x_3)\, dx_1\, dx_2\right\} \quad \text{say.}$$

The integrals appearing in the last line (and which are defined in an obvious way by the equality relation) are also known as *surface integrals* and in some treatments of the topic the more general surface integral is built up from integrals of this type.

(iii) It will be observed that the *area* of S (p. 181) is given by

$$\iint_S \mathbf{n}.\mathbf{n}\,ds = \iint_S ds.$$

(iv) It is conventional in the case of closed surfaces to select the outward-pointing normal $\mathbf{n}$ or in the case of oriented surfaces to select that normal which in conjunction with the orientation of the surface forms a right-handed system.

Example. Let $\mathbf{f}: \mathbb{R}^3 \to \mathbb{R}^3$ be defined by $(x, y, z) \mapsto x^3\mathbf{i}+y^3\mathbf{j}+z^3\mathbf{k}$. Then if S is the hemisphere defined in the example on p. 180,

$$\int_S \mathbf{f}.\mathbf{n}\,ds = \int_S \mathbf{f}.\mathbf{ds} = \int_R (\sin^3 t_1 \cos^3 t_2 \mathbf{i}+\sin^3 t_1 \sin^3 t_2 \mathbf{j}+\cos^3 t_1 \mathbf{k})$$

$$.\,(\sin^2 t_1 \cos t_2 \mathbf{i}+\sin^2 t_1 \sin t_2 \mathbf{j}+\sin t_1 \cos t_1 \mathbf{k})\,d(t_1, t_2)$$

$$= \int_0^{2\pi}\left[\int_0^{\frac{1}{2}\pi}(\sin^5 t_1 \cos^4 t_2+\sin^5 t_1 \sin^4 t_2+\sin t_1 \cos^4 t_1)\,dt_1\right]dt_2$$

$$= \int_0^{2\pi}\left[\int_0^1 ((1-u^2)^2(\cos^4 t_2+\sin^4 t_2)+u^4)\,du\right]dt_2$$

$$= \int_0^{2\pi}[\tfrac{8}{15}(\cos^4 t_2+\sin^4 t_2)+\tfrac{1}{5}]\,dt_2$$

$$= \frac{2\pi}{5}+\frac{16}{15}\int_0^{2\pi}\cos^4 t_2\,dt_2$$

$$= \frac{2\pi}{5}+\frac{16}{15}\int_0^{2\pi}\left[\frac{1}{4}\left(\frac{\cos 4t_2+1}{2}\right)+\tfrac{1}{2}\cos 2t_2+\tfrac{1}{4}\right]dt_2$$

$$= \frac{2\pi}{5}+\frac{16}{15}.2\pi(\tfrac{1}{8}+\tfrac{1}{4})$$

$$= \frac{2\pi}{15}(3+16.\tfrac{3}{8}) = \frac{6\pi}{5}.$$

An alternative presentation of the integral is

$$\iint_S (x^3\,dy\,dz+y^3\,dz\,dx+z^3\,dx\,dy) = \tfrac{6}{5}\pi.$$

See also: Gauss's (Divergence) Theorem (p. 203); Green's Theorem (p. 204); Stokes's Theorem (p. 210).

37 Measure and Lebesgue integration

A function, $\mu : \mathscr{C} \to \overline{\mathbb{R}}$, whose domain is a non-empty class $\mathscr{C}$ of subsets of some set X (assumed to include the empty set $\varnothing$) and whose codomain is the extended real number system (p. 67), is said to be a **finitely additive set function** if $\mu(\varnothing) = 0$ and if for every collection $E_1, \ldots, E_n$ of pairwise disjoint sets of $\mathscr{C}$ such that their union

$$\bigcup_{i=1}^{n} E_i$$

belongs to $\mathscr{C}$ we have $\mu\left(\bigcup_{i=1}^{n} E_i\right) = \sum_{i=1}^{n} \mu(E_i)$.

Note. (i) The condition $\mu(\varnothing) = 0$ is redundant provided $\operatorname{Im}(\mu) \not\subset \{-\infty, +\infty\}$ and it is normal to assume that $\operatorname{Im}(\mu)$ does not contain both $-\infty$ and $+\infty$.

(ii) μ will be finitely additive on a ring of sets, $\mathscr{R}$ (p. 78), if and only if $\mu(\varnothing) = 0$ (see note (i)) and

$$(E, F \in \mathscr{R}, E \cap F = \varnothing) \Rightarrow \mu(E \cup F) = \mu(E) + \mu(F).$$

(iii) If μ is finitely additive on a ring of sets $\mathscr{R}$, then

$$\mu(E) + \mu(F) = \mu(E \cup F) + \mu(E \cap F) \quad (\text{all } E, F \in \mathscr{R}).$$

In general, for all $E_1, \ldots, E_n \in \mathscr{R}$ we have

$$\mu\left(\bigcup_{i=1}^{n} E_i\right) = \sum_{i=1}^{n} \mu(E_i) - \sum_{i<j}^{n} \mu(E_i \cap E_j) + \sum_{i<j<k}^{n} \mu(E_i \cap E_j \cap E_k) - \ldots + (-1)^{n+1} \mu(E_1 \cap E_2 \cap \ldots \cap E_n),$$

a formula known as the **inclusion and exclusion formula.**

An additive set function μ is said to be **countably additive, completely additive,** or **σ-additive** if for every countable collection $E_1, E_2, \ldots$ of pairwise disjoint sets of $\mathscr{C}$ whose union is in $\mathscr{C}$, we have

$$\mu\left(\bigcup_{i=1}^{\infty} E_i\right) = \sum_{i=1}^{\infty} \mu(E_i).$$

A **measure** is defined to be a non-negative, countably additive set function (i.e. a σ-additive set function with codomain $\{x \in \overline{\mathbb{R}} \mid x \geqslant 0\}$).

If μ is a measure on a ring $\mathscr{R}$, then a set $E \in \mathscr{R}$ is said to have **finite measure** if $\mu(E) \in \mathbb{R}$. E is said to have **σ-finite measure** if there is a sequence (E_n) of sets of finite measure in $\mathscr{R}$ such that

$$E \subset \bigcup_{i=1}^{\infty} E_i.$$

If every set in $\mathscr{R}$ has finite (σ-finite) measure, then μ is said to be **finite** (or **σ-finite**) on $\mathscr{R}$. If $\mathscr{R}$ is an algebra of sets (p. 78), then μ is called **totally finite (totally σ-finite)**.

Given two classes $\mathscr{C}$ and $\mathscr{D}$ of subsets of X satisfying $\mathscr{C} \subset \mathscr{D}$, and two set functions μ, ν defined upon $\mathscr{C}$ and $\mathscr{D}$ respectively, we say that ν is an **extension** of μ and that μ is a **restriction** of ν if for all $E \in \mathscr{C}$ we have

$$\mu(E) = \nu(E).$$

A non-empty class $\mathscr{C}$ of sets is said to be **hereditary** if

$$(E \in \mathscr{C} \text{ and } F \subset E) \Rightarrow F \in \mathscr{C}. \quad \text{(See also note (i) below.)}$$

A set function $\mu^* : \mathscr{C} \to \mathbb{R}$ is said to be **monotone** if

$$(E, F \in \mathscr{C} \text{ and } E \subset F) \Rightarrow \mu^*(E) \leqslant \mu^*(F).$$

A non-negative set function $\mu^* : \mathscr{C} \to \{x \in \mathbb{R} | x \geqslant 0\}$ is said to be **subadditive** if

$$(E, F, E \cup F \in \mathscr{C}) \Rightarrow \mu^*(E \cup F) \leqslant \mu^*(E) + \mu^*(F).$$

A subadditive set function μ^* is **countably subadditive** if, for every sequence (E_i) of sets of $\mathscr{C}$ whose union is also in $\mathscr{C}$,

$$\mu^*\left(\bigcup_{i=1}^{\infty} E_i\right) \leqslant \sum_{i=1}^{\infty} \mu^*(E_i).$$

A monotone, countably subadditive set function μ^* defined on a hereditary σ-ring $\mathscr{H}$ and such that $\mu^*(\varnothing) = 0$ is known as an **outer measure**.

Note. (i) The class of all subsets of some set X is a hereditary σ-ring and some authors restrict the definition of an outer measure to such classes.

(ii) Every measure defined on a hereditary ring is also an outer measure.

Examples. (i) Let $\mathscr{C} = \mathscr{P}(\{1, 2, 3\})$. Then $\mu : \mathscr{C} \to \mathbb{R}$ defined by

$$\begin{aligned} \mu(\varnothing) &= 0, \\ \mu(\{1\}) = \mu(\{2\}) &= \mu(\{3\}) = \tfrac{1}{3}, \\ \mu(\{1, 2\}) = \mu(\{2, 3\}), &= \mu(\{1, 3\}) = \tfrac{2}{3}, \\ \mu(\{1, 2, 3\}) &= 1, \end{aligned}$$

is a *finite measure*. (This measure can be associated with the probabilities assigned to three distinct 'equally likely' events 1, 2 and 3 and to combinations of these events.) μ is readily seen to be *monotone* and *subadditive*.

(ii) Let $\mathscr{C} = \mathscr{P}(\mathbb{R})$. Then $\mathscr{C}$ is a *hereditary σ-ring* and $\mu^* : \mathscr{C} \to \{0, 1\}$ defined by

$$\mu^*(E) = 0 \quad \text{if } E \text{ is countable}$$
$$\mu^*(E) = 1 \quad \text{if } E \text{ is not countable,}$$

is an *outer measure* on $\mathscr{C}$. μ^* is *not* a measure since it is not finitely additive.

Lebesgue measure

We define an **interval** in $\mathbb{R}^n$ to be a set of the form

$$\{(x_1, x_2, \ldots, x_n) | a_i * x \dagger b_i, i = 1, 2, \ldots, n\}$$

where $*$ and $\dagger$ denote either $<$ or $\leqslant$. We allow the possibility that $a_i = b_i$ and include the empty set amongst the set of intervals.

If A is the union of a finite number of such intervals, we say that it is an **elementary set**.

Note. Some authors define an *elementary set* (or *figure*) to be the union of a finite number of half-open intervals, i.e. intervals obtained as the Cartesian product of n half-open intervals in $\mathbb{R}$. This will affect the proofs required but not the general results.

The class of all elementary sets is a ring, which we denote by $\mathscr{E}^n$, and it can be shown that if $A \in \mathscr{E}^n$ then A can be presented as the finite union of pairwise disjoint intervals

$$A = I_1 \cup I_2 \cup \ldots \cup I_N.$$

If then we denote the content (p. 159) of an interval I by

$$m(I) \left(= \prod_{i=1}^{n} (b_i - a_i) \right),$$

we can define a non-negative, additive set function $m : \mathscr{E}^n \to \overline{\mathbb{R}}$, by

$$m(A) = m(I_1) + m(I_2) + \ldots + m(I_N).$$

In the cases $n = 1$, 2 and 3, m is length, area and volume respectively.

$\mathscr{E}^n$ is not a σ-ring and does not contain all those subsets of $\mathbb{R}^n$ in which we are interested. We attempt therefore to obtain an extension of m to a larger class of sets.

Consider countable coverings of any set $E \subset \mathbb{R}^n$ by open sets of $\mathscr{E}^n$,

$$E \subset \bigcup_{i=1}^{\infty} A_i, \quad \text{say.}$$

Such a covering, i.e. by a countable number of open sets, is called a **Lebesgue covering**.

Define $\overline{m}(E)$ to be glb $\sum_{i=1}^{\infty} m(A_i)$ when the glb is taken over all Lebesgue coverings of E.

It can then be shown that $\overline{m}$, as defined above, is an extension of m from $\mathscr{E}^n$ to the class of all subsets of $\mathbb{R}^n$ and that it is an outer measure on that σ-ring. $\overline{m}$ is known as the **Lebesgue outer measure.**

Given $A, B \subset \mathbb{R}^n$ we define

$$\mathrm{d}(A, B) = \overline{m}(A \,\Delta\, B)$$

(Δ denotes the symmetric difference (p. 10)), and write $A_i \to A$ if $A, A_i \subset \mathbb{R}^n$ ($i = 1, 2, \ldots$) and

$$\lim_{i\to\infty} \mathrm{d}(A, A_i) = 0.$$

If there is a sequence of sets of $\mathscr{E}^n$ such that $A_i \to A$, we say that A is **finitely *m*-measurable** and write $A \in \mathscr{M}_F(m)$.

If A is the union of a countable collection of finitely m-measurable sets, then A is said to be ***m*-measurable** and we write $A \in \mathscr{M}(m)$.

It can be shown that $\mathscr{M}(m)$ is a σ-ring and that, setting $m(A) = \overline{m}(A)$ for all $A \in \mathscr{M}(m)$, we obtain an extension of m from $\mathscr{E}^n$ to the class $\mathscr{M}(m)$. m is known as the **Lebesgue measure** on $\mathbb{R}^n$ and we say that an m-measurable set is **Lebesgue measurable.**

Note. (i) All open sets of $\mathbb{R}^n$ and all closed sets of $\mathbb{R}^n$ belong to $\mathscr{M}(m)$. It follows that all Borel sets (p. 107) in $\mathbb{R}^n$ are Lebesgue measurable.

(ii) Not all subsets of $\mathbb{R}^n$ are Lebesgue measurable. This can be shown as follows. Define an equivalence relation on $\mathbb{R}$ by xEy if $x - y = n + m\pi$ for some $n, m \in \mathbb{Z}$. E will partition $\mathbb{R}$ into disjoint cosets and it can be shown that the set T constructed (making use of the axiom of choice (p. 201)) by taking exactly one point from each coset is not Lebesgue measurable.

A set E for which $\overline{m}(E) = 0$ is said to be a set of (Lebesgue) **measure zero.**

Note. (i) Every countable set has Lebesgue measure zero. The converse, however, is not true, see, for example, the Cantor set (p. 199).

(ii) The sets of measure zero form a σ-ring.

(iii) The term '**almost everywhere**' (a.e.) is used to mean 'except on a set of (Lebesgue) measure zero'.

(iv) It can be shown that every Lebesgue measurable set in $\mathbb{R}^n$ can be presented as the union of a Borel set and a set of measure zero.

(v) All sets of Lebesgue measure zero are Lebesgue measurable.

Measurable functions

We say that a set X is a **measurable space** (X, S) if there is a σ-ring S of subsets of X such that $\cup S = X$. A **measure space**, (X, S, μ), is a measurable space together with a measure μ defined on S (the σ-ring of measurable sets).

Thus $(\mathbb{R}^n, \mathcal{M}(m), m)$ is a measure space.

Let $f: X \to \overline{\mathbb{R}}$ be a function defined on a measure space X. We say that f is **measurable** if the set

$$\{x \mid f(x) > a\}$$

is measurable for all $a \in \mathbb{R}$.

Note. (i) In the above definition the symbol '$>$' can be replaced by '$\geqslant$', '$\leqslant$' or '$<$' without affecting the resulting theory.

(ii) It can be shown that measurable functions possess many properties such as: the sum and product of measurable functions are measurable, and the limit of a convergent sequence of measurable functions (p. 120) is measurable.

A function $f: X \to \mathbb{R}$ (note the restricted codomain) is said to be **simple** if there is a finite, pairwise disjoint class $\{E_1, \ldots, E_n\}$ of measurable sets and a finite set $\{\alpha_1, \ldots, \alpha_n\}$ of real numbers such that

$$f(x) = \begin{cases} \alpha_i & \text{if } \ x \in E_i \ \ (i = 1, \ldots, n), \\ 0 & \text{if } \ x \notin \bigcup_{i=1}^{n} E_i. \end{cases}$$

An important example of a simple function is the **characteristic function** χ_E (or K_E) of a measurable set E defined by

$$\chi_E(x) = \begin{cases} 1 & \text{if } \ x \in E, \\ 0 & \text{if } \ x \notin E. \end{cases}$$

A simple function can, therefore, be presented in the form

$$f = \sum_{i=1}^{n} \alpha_i \chi_{E_i}.$$

It can be shown that every measurable function is the limit of a sequence of simple functions and that if the function is non-negative, then the sequence can be chosen to be monotonically increasing.

Integration

A simple function $$f = \sum_{i=1}^{n} \alpha_i \chi_{E_i}$$

on a measure space (X, S, μ) is **integrable** if $\mu(E_i)$ is finite for every i for which α_i is non-zero.

The integral of f, written

$$\int f \, d\mu,$$

is defined to be $\sum_{i=1}^{n} \alpha_i \mu(E_i)$ (where $0 \times (+\infty) = 0$).

Note. If f is integrable, the value of the integral is finite.

If E is a measurable set and f is an integrable, simple function, then we define the **integral of f over E**, written

$$\int_E f \, d\mu,$$

to be the integral of the product function $\chi_E f$.

If f is a measurable, non-negative function we define the **integral of f over E** with respect to the measure μ, denoted by

$$\int_E f \, d\mu,$$

to be $$\text{lub}\left\{\int_E s \, d\mu\right\}$$

where the lub is taken over all simple functions s such that $0 \leqslant s \leqslant f$. f is said to be **integrable over E** if the integral is finite.

Alternatively, we can define

$$\int_E f \, d\mu = \lim_{n \to \infty} \int_E f_n \, d\mu$$

where (f_n) is a monotone increasing sequence of simple functions such that $f_n \to f$ (see p. 188). This limit is independent of the choice of sequence.

If f is any measurable function, then the functions $f^+ = \max(f, 0)$ and $f^- = -\min(f, 0)$ are also measurable and are non-negative. We can, therefore, form

$$\int_E f^+ d\mu \quad \text{and} \quad \int_E f^- d\mu.$$

If at least one of these integrals is finite we define the **integral of f over E**, written $\int_E f\,d\mu$, to be

$$\int_E f^+\,d\mu - \int_E f^-\,d\mu.$$

If $\int_E f\,d\mu$ is finite, we say that f is **integrable** (or **summable**) **over E**.

Note. (i) If f is measurable and bounded on E and if $\mu(E) < \infty$, then f is integrable over E.

(ii) If E has measure zero (i.e. if $\mu(E) = 0$) and f is measurable, then

$$\int_E f\,d\mu = 0.$$

(iii) If f is integrable over E and A is a measurable subset of E, then f is integrable over A.

(iv) If B is a subset of a measurable set A and $A-B$ is a set of measure zero, then

$$\int_A f\,d\mu = \int_B f\,d\mu$$

whenever the integrals exist.

Again, if f and g are two measurable functions which are equal almost everywhere (with respect to μ) (p. 187) on a measurable set A, then

$$\int_A f\,d\mu = \int_A g\,d\mu,$$

provided the integrals exist.

In the special case when $X = \mathbb{R}^n$, μ is the Lebesgue measure and S is the set of Lebesgue measurable functions, then it is usual to denote the integral – the **Lebesgue integral** – by

$$\int_E f(\mathbf{x})\,d\mathbf{x} \quad \text{or} \quad \int\!\!\int \dots \int_E f(x_1, \dots, x_n)\,dx_1 \dots dx_n$$

rather than $\int_E f\,dm$.

Note. (i) It is a consequence of (iv) above that, say, if E is an interval in $\mathbb{R}$ with end-points a, b then the notation

$$\int_a^b f(x)\,dx$$

can be used without ambiguity to denote the Lebesgue integral, the value of which will not depend on whether or not E is open, closed or half-open.

(ii) Since infinite intervals are measurable sets, we do not (as is the case with Riemann integrals (p. 146)) have to define the integral over such an interval as the limit of integrals over finite intervals.

(iii) Every function which is Riemann integrable is Lebesgue integrable and the values of the two integrals are equal.

Example. To evaluate the Lebesgue integral

$$\int_0^1 x^2\,dx.$$

Let E be the set $\langle 0, 1\rangle$. Then E is an *elementary set* having Lebesgue measure 1.

The function $f: \mathbb{R} \to \mathbb{R}$ defined by

$$f(x) = x^2 \quad \text{if} \quad x \in E,$$
$$f(x) = 0 \quad \text{if} \quad x \notin E$$

is a measurable function since $\{x \mid f(x) \geqslant a\}$ is measurable for all $a \in \mathbb{R}$ (indeed $\{x \mid f(x) \geqslant a\}$, depending on whether $a \leqslant 0$, $0 < a < 1$, or $a \geqslant 1$, is $\mathbb{R}$, $\langle\sqrt{a}, 1\rangle$ or $\varnothing$).

We define a *monotonic increasing sequence of simple functions* tending to f as follows.

Let $$Q_{p,s} = \left\{x \,\middle|\, \frac{p-1}{2^s} \leqslant x < \frac{p}{2^s}\right\} \quad (p = 1, \ldots, 2^s,\ s = 1, 2, 3, \ldots)$$

and define $$f_s(x) = \left(\frac{p-1}{2^s}\right)^2 \quad \text{for} \quad x \in Q_{p,s} \quad (p = 1, \ldots, 2^s),$$

$$= 0 \quad \text{for} \quad x \in \mathbb{R} - E.$$

Then, for all $x \in \mathbb{R}$, $\quad 0 \leqslant f_s(x) \leqslant f(x).$

Moreover, for all $x \in \mathbb{R}$, we have

$$f_{s+1}(x) \geqslant f_s(x),$$

i.e. (f_s) is a monotonic increasing sequence, and

$$0 \leqslant f(x) - f_s(x) \leqslant \left(\frac{2^s}{2^s}\right)^2 - \left(\frac{2^s-1}{2^s}\right)^2 = \frac{2.2^s-1}{2^s.2^s} < \frac{2}{2^s},$$

so that $f_s(x) \to f(x)$ as $s \to \infty$.

Now $$\int_E f_s\,dx = \frac{1}{2^s}\left[0 + \left(\frac{1}{2^s}\right)^2 + \left(\frac{2}{2^s}\right)^2 + \ldots + \left(\frac{2^s-1}{2^s}\right)^2\right]$$

$$= \frac{1}{n^3}\,[1 + 2^2 + 3^2 + \ldots + (n-1)^2] \quad (\text{setting } n = 2^s)$$

$$= \frac{1}{n^3}\left[\frac{(n-1)n(2n-1)}{6}\right] \to \frac{1}{3} \quad \text{as} \quad s \to \infty.$$

It follows then that $$\int_0^1 x^2\,dx = \tfrac{1}{3}.$$

38 Fourier series

Although later in this section we shall be concerned mainly with real-valued functions of a real variable, the opening definitions are given for more general complex-valued functions since the extra generality is obtained without much additional complication.

Let f and g be two complex-valued functions defined and Riemann-integrable on an interval $\leqslant a, b \geqslant$ of $\mathbb{R}$. We define the **inner product** of f and g, denoted by $\langle f, g \rangle$, to be

$$\int_a^b f(x)\, \overline{g(x)}\, \mathrm{d}x.$$

The non-negative number $\sqrt{\langle f, f \rangle}$, denoted by $\|f\|$, is known as the **norm** of f.

Note. (i) The inner product has the following properties

(*a*) $\langle f, g \rangle = \overline{\langle g, f \rangle}$,

(*b*) $\langle \lambda f_1 + \mu f_2, g \rangle = \lambda \langle f_1, g \rangle + \mu \langle f_2, g \rangle \quad (\lambda, \mu \in \mathbb{C})$.

It will be observed that these properties correspond to those of an inner product on a *finite-dimensional* vector space over $\mathbb{C}$, for example, the inner product is not symmetrical (see p. 92 and in particular, note (ii)).

(ii) The integral corresponds to the sum

$$\sum_{i=1}^{n} x_i \overline{y_i}$$

which defines the inner product (scalar product) of two vectors $\mathbf{x}$ and $\mathbf{y}$ in a vector space of dimension n over $\mathbb{C}$ (p. 92). The norm of f corresponds to the length (norm) of the vector $\mathbf{x}$. Below we give definitions analogous to those given on p. 90 for **orthogonal** and **orthonormal** bases.

Let (ϕ_n) $(n = 0, 1, 2, \ldots)$ be a sequence of complex-valued functions defined on $\leqslant a, b \geqslant$ such that

$$\langle \phi_n, \phi_m \rangle = 0 \quad \text{whenever} \quad m \neq n.$$

Then (ϕ_n) is said to be an **orthogonal system** of functions on $\leqslant a, b \geqslant$. If, in addition, $\langle \phi_n, \phi_n \rangle = 1$ for all n, then (ϕ_n) is said to be **orthonormal**.

Examples. (i) *Orthonormal systems* of particular importance are those on $\leqslant a, a+2\pi \geqslant$ given by

$$(a)\ \phi_0(x) = \frac{1}{\sqrt{2\pi}},\quad \phi_{2n-1}(x) = \frac{\cos nx}{\sqrt{\pi}},\quad \phi_{2n}(x) = \frac{\sin nx}{\sqrt{\pi}}\quad (n = 1, 2, 3, \ldots),$$

$$(b)\ \phi_n(x) = \frac{1}{\sqrt{2\pi}}\, e^{inx}\quad (n \in \mathbb{N}).$$

We shall refer to these systems as F_1 and F_2 respectively.

(ii) Another *orthogonal system* of importance is the sequence of **Legendre polynomials** $(\phi_0, \phi_1, \phi_2, \ldots)$, defined by

$$\phi_0(x) = 1,\quad \phi_1(x) = x,\quad \phi_2(x) = \tfrac{3}{2}x^2 - \tfrac{1}{2}, \ldots,$$

$$\phi_n(x) = \frac{1}{2^n n!}\frac{d^n}{dx^n}(x^2 - 1)^n.$$

This set of functions is orthogonal on $\leqslant -1, 1 \geqslant$.

The *finite* set of functions $\{\phi_0, \phi_1, \ldots, \phi_n\}$ is said to be **linearly dependent** on $\leqslant a, b \geqslant$ whenever constants $c_0, c_1, \ldots, c_n \in \mathbb{C}$, not all of which are zero, can be found such that

$$c_0\phi_0(x) + c_1\phi_1(x) + \ldots + c_n\phi_n(x) = 0 \quad \text{for all} \quad x \in \leqslant a, b \geqslant.$$

A *finite* set which is not linearly dependent is said to be **linearly independent**. An *infinite* sequence of functions is said to be **linearly independent** on $\leqslant a, b \geqslant$ if every finite subsequence is linearly independent.

All orthonormal systems are linearly independent.

If (ϕ_n) is orthonormal on $\leqslant a, b \geqslant$ and if f is defined and Riemann-integrable on $\leqslant a, b \geqslant$, then we call

$$c_n = \int_a^b f(x)\overline{\phi_n(x)}\, dx \quad (n \in \mathbb{N})$$

the nth **Fourier coefficient of f relative to (ϕ_n).**

We write

$$f(x) \sim \sum_{n=0}^{\infty} c_n\phi_n(x)$$

and call this series the **Fourier series of f relative to (ϕ_n).**

Note that the use of the symbol $\sim$ is not to be taken to imply that the series is convergent or that, if it is convergent, it converges to $f(x)$.

If f is a real-valued function defined and Riemann-integrable on $\leqslant a, a+2\pi \geqslant$ and (ϕ_n) is the sequence F_1 defined above, then we write

$$f(x) \sim \frac{a_0}{2} + \sum_{n=1}^{\infty} (a_n \cos nx + b_n \sin nx)$$

where
$$a_n = \frac{1}{\pi}\int_a^{a+2\pi} f(t)\cos nt\,dt, \quad b_n = \frac{1}{\pi}\int_a^{a+2\pi} f(t)\sin nt\,dt,$$

and refer to the series as *the* **Fourier series generated by** f or as *the* **trigonometric Fourier series** for f.

In many applications f is a periodic function defined on $\mathbb{R}$ with period 2π and a, the end-point of the interval of integration, can be chosen at will. An obvious choice is to set $a = 0$, and integrate over the interval $\langle 0, 2\pi \rangle$. Use of the interval $\langle -\pi, \pi \rangle$ can, however, facilitate the calculation of the Fourier coefficients of functions which are odd or even (p. 120).

Example. Consider the function $f: \mathbb{R} \to \mathbb{R}$ defined by

$$f(x) = x \qquad (0 \leqslant x < 2\pi).$$

$$f(x+2\pi) = f(x) \qquad (\text{all } x \in \mathbb{R}).$$

Then the *Fourier series generated by f* has *coefficients*

$$a_n = \frac{1}{\pi}\int_0^{2\pi} t\cos nt\,dt, \quad b_n = \frac{1}{\pi}\int_0^{2\pi} t\sin nt\,dt.$$

That is, $a_n = \frac{1}{\pi}\left(\left[\frac{t\sin nt}{n}\right]_0^{2\pi} - \int_0^{2\pi} \frac{\sin nt}{n}\,dt\right)$ (provided $n \neq 0$)

$$= +\frac{1}{\pi}\left[\frac{\cos nt}{n^2}\right]_0^{2\pi} = 0.$$

When $n = 0$, $a_0 = \frac{1}{\pi}\int_0^{2\pi} t\,dt = \frac{1}{\pi}\left[\frac{t^2}{2}\right]_0^{2\pi} = 2\pi.$

$$b_n = \frac{1}{\pi}\left(\left[-t\frac{\cos nt}{n}\right]_0^{2\pi} + \int_0^{2\pi} \frac{\cos nt}{n}\,dt\right)$$

$$= \frac{1}{\pi}\left(\frac{-2\pi}{n}\right) = \frac{-2}{n}.$$

Hence, the *Fourier series* for f is

$$\pi - 2\sum_{n=1}^{\infty} \frac{\sin nx}{n}.$$

(The Fourier series does, in fact, converge to $f(x)$ on $\mathbb{R} - \{2n\pi\}$, $n \in \mathbb{Z}$.)

Various theorems can be proved concerning Fourier series of which the following are amongst the most important:

If (ϕ_n) is orthonormal on $\langle a, b \rangle$, f is a real-valued Riemann-integrable function on $\langle a, b \rangle$ and c_n are the Fourier coefficients of f relative to (ϕ_n), then

(*a*) $c_n \to 0$ as $n \to \infty$;

(*b*) $\sum_{n=0}^{\infty} |c_n|^2$ converges and satisfies the inequality (known as **Bessel's inequality**)

$$\sum_{n=0}^{\infty} |c_n|^2 \leqslant \int_a^b |f(x)|^2 \, dx;$$

(*c*) $\sum_{n=0}^{\infty} |c_n|^2 = \int_a^b |f(x)|^2 \, dx$ **(Parseval's equation)**,

if and only if $$\lim_{n \to \infty} \left\| f - \sum_{i=0}^{n} c_i \phi_i \right\| = 0.$$

Note. (i) Setting $(\phi_n) = F_2$ (p. 193), then it follows from (*a*) that

$$\lim_{n \to \infty} \int_{-\pi}^{\pi} f(x) \cos nx \, dx = \lim_{n \to \infty} \int_{-\pi}^{\pi} f(x) \sin nx \, dx = 0.$$

(ii) If Parseval's equation is written in the form

$$\|f\|^2 = c_0^2 + c_1^2 + c_2^2 + \ldots,$$

then we obtain a formula similar to

$$|\mathbf{x}|^2 = x_1^2 + x_2^2 + \ldots + x_n^2.$$

If Parseval's equation holds for every function belonging to a set S of Riemann-integrable functions defined on $\langle a, b \rangle$, then the sequence (ϕ_n) is said to be **complete** for S.

Note. If (ϕ_n) is complete, then one cannot add an integrable function f to the set $\{\phi_n\}$ so as to form a 'larger' orthonormal set.

The trigonometric sequence F_1 (p. 193) can be shown to be complete for the set of all functions which are continuous on a closed interval of length 2π.

The determining of necessary and sufficient conditions for the trigonometric Fourier series to converge to the value of the function at every point of $\langle -\pi, \pi \rangle$ is a problem which still awaits solution. One example of the difficulties associated with the problem is that two functions whose values differ at only a finite number of points will have the same Fourier coefficients and, hence, the same Fourier series.

However, many important results have been obtained such as **Fejér's Theorem** which states that if f is continuous and periodic with period 2π, then the Fourier series generated by f is Cesàro summable (p. 142) and has $(C, 1)$ sum $f(x)$. (The sequence of arithmetic means of the partial sums of the Fourier series of f tends to $f(x)$ *uniformly* for all x.)

It is a consequence of Fejér's Theorem that if two continuous functions f and g have the same Fourier series, then $f(x) = g(x)$ for all x.

Functions with periods other than 2π

If f is Riemann integrable on $\leqslant 0, p \geqslant$ and has period p (p. 120) then the trigonometric Fourier series for f is written

$$\frac{a_0}{2} + \sum_{n=1}^{\infty}\left(a_n \cos\frac{2\pi nx}{p} + b_n \sin\frac{2\pi nx}{p}\right)$$

where the Fourier coefficients are given by

$$a_n = \frac{2}{p}\int_0^p f(t)\cos\frac{2\pi nt}{p}\,dt,$$

$$b_n = \frac{2}{p}\int_0^p f(t)\sin\frac{2\pi nt}{p}\,dt \quad (n = 0, 1, 2, \ldots).$$

Exponential forms. Making use of the formula

$$e^{inx} = \cos nx + i\sin nx$$

(p. 203) the Fourier series for a function with period p can be written

$$\sum_{-\infty}^{\infty} \alpha_n e^{2\pi inx/p}$$

where
$$\alpha_n = \frac{1}{p}\int_0^p f(t)\,e^{-2\pi int/p}\,dt.$$

Example. Let $f: \mathbb{R} \to \mathbb{R}$ be defined by

$$f(x) = 1 \text{ if } x \in \,<2n, 2n+1> \quad (n \in \mathbb{Z}),$$

$$f(x) = -1 \text{ if } x \in \,<2n-1, 2n> \quad (n \in \mathbb{Z}),$$

$$f(x) = 0 \text{ if } x \in \mathbb{Z}.$$

Then f is *periodic* with *period* 2.

Hence, the *Fourier coefficients* for f are given by

$$a_0 = \int_0^1 dt - \int_1^2 dt = 0,$$

$$a_n = \left(\int_0^1 \cos n\pi t\,dt - \int_1^2 \cos n\pi t\,dt\right) = 0 \quad (n \neq 0),$$

$$b_n = \left(\int_0^1 \sin n\pi t\,dt - \int_1^2 \sin n\pi t\,dt\right)$$

$$= \frac{1}{n\pi}(1 - 2\cos n\pi + \cos 2n\pi)$$

$$= \frac{2}{n\pi}(1 - \cos n\pi).$$

Hence,

$$b_{2n} = 0, \quad b_{2n-1} = \frac{4}{(2n-1)\pi},$$

and the *Fourier series* for f is

$$\sum_{n=1}^{\infty} \frac{4 \sin (2n-1)\pi x}{(2n-1)\pi}.$$

Appendix 1 Some 'named' theorems and properties

Abel, Niels Henrik (1802–29)
Abelian group p. 26.
A's Limit Theorem Suppose that $\Sigma a_n x^n$ has radius of convergence r and that $\Sigma a_n r^n$ is convergent. Then

$$\lim_{x \to r^-} \Sigma a_n x^n = \Sigma a_n r^n = \Sigma(\lim_{x \to r^-} a_n x^n).$$

A's test for convergence The series $\Sigma a_n b_n$ converges if Σa_n converges and if (b_n) is a monotonic convergent sequence.
Abelian Groups, Fundamental Theorem of Every finitely generated Abelian group is the direct sum (p.64) of cyclic groups.
Algebra, Fundamental Theorem of Every non-constant polynomial with coefficients in $\mathbb{C}$ has a root in $\mathbb{C}$.

Argand, Jean Robert (1768–1822)
A diagram p. 68.

Banach, Stefan (1892–1945)
B space p. 111.

Bernstein, Felix (1878–1956)
See Schröder (p. 210).

Bessel, Friedrich Wilhelm (1784–1846)
B's inequality p. 195.

Bolzano, Bernard (1781–1848)
B's Theorem (*Intermediate Value Theorem*) Let f be real valued and continuous on $\leqslant a, b \geqslant \subset \mathbb{R}$ and suppose that $f(a)$ and $f(b)$ have different signs. Then there is at least one point $x \in \langle a, b \rangle$ such that $f(x) = 0$.

B–Weierstrass Theorem A bounded infinite sequence of real numbers always contains convergent subsequences. (More generally: if a bounded set S in $\mathbb{R}^n$ contains infinitely many points, then there exists a point $x \in \mathbb{R}^n$ which is a limit point (p. 101) of S.)

Boole, George (1815–64)
B'ean algebra p. 76, **field** p. 78, **lattice** p. 76, **ring** p. 78.

Borel, Emile (1871–1956)
B field p. 79, **set** p. 107.
B'ian set p. 107. *See* Heine (p. 204).

Burnside, William (1852–1927)
B problem If a group G is generated by a finite number of elements and if the least common multiple of the orders of the elements of G is finite, is G a finite group?

Calculus, Fundamental Theorem of the If f is Riemann-integrable on $\leqslant a, b \geqslant$ and if there is a differentiable function F on $\leqslant a, b \geqslant$ such that $F'(x) = f(x)$ for all $x \in \leqslant a, b \geqslant$, then

$$\int_a^b f(x)\,\mathrm{d}x = F(b) - F(a).$$

(A more general version of this theorem is given on p. 136.)

Cantor, Georg (1845–1918)
C–Bernstein Theorem *See* Schröder (p. 210).
C fundamental sequence p. 113.
C set Denote the open interval $<(3r-2)/3^n, (3r-1)/3^n>$ by $I_{n,r}$ and put

$$G_n = \bigcup_{r=1}^{3^{n-1}} I_{n,r}, \quad G = \bigcup_{n=1}^{\infty} G_n.$$

The set $C = \leqslant 0, 1 \geqslant - G$ is called the Cantor ternary set. It has cardinal **c** (p. 21), is perfect (p. 103), nowhere dense (p. 103) and has Lebesgue measure zero (p. 187).

Cartesian *See* Descartes (p. 202).

Cauchy, Augustin Louis (1789–1857)
C principal value p. 147, **product** p. 141, **remainder** p. 212, **sequence** p. 109. *See* Schwarz (p. 210).
C's condensation test If Σa_n is a series of monotonic decreasing, positive, real terms, then the series

$$\Sigma a_n \quad \text{and} \quad \Sigma h^n a_{h^n} \quad (h = 2, 3, \ldots)$$

converge or diverge together.
C condition for convergence p. 115.
C integral formula Let f be analytic (p. 154) on an open region (p. 152) S of $\mathbb{R}^2$ and $\boldsymbol{\Gamma}$ be a rectifiable Jordan curve (p. 151) such that both $\boldsymbol{\Gamma}$ and its interior lie entirely within S. Then, provided $\boldsymbol{\Gamma}$ is positively oriented, we have

$$f(z_0) = \frac{1}{2\pi \mathrm{i}} \int_{\boldsymbol{\Gamma}} \frac{f(z)}{z - z_0}\,\mathrm{d}z$$

(p. 157) for every point z_0 within $\boldsymbol{\Gamma}$.
C Integral Theorem Let f be analytic (p. 154) on an open region (p. 152) S of $\mathbb{R}^2$ and $\boldsymbol{\Gamma}$ be a rectifiable Jordan curve (p. 151) such that both $\boldsymbol{\Gamma}$ and its interior lie entirely within S. Then

$$\int_{\boldsymbol{\Gamma}} f(z)\,\mathrm{d}z = 0.$$

C Residue Theorem Let f be analytic (p. 154) at all points of an open region (p. 152) S of $\mathbb{R}^2$ with the exception of a finite number of isolated singularities (p. 155). Let $\boldsymbol{\Gamma}$ be a rectifiable Jordan curve (p. 151) such that both it and its interior lie within S. If f has no singularities on $\boldsymbol{\Gamma}$ and its singularities interior to $\boldsymbol{\Gamma}$ are $z_1, z_2, \ldots, z_n$, then, provided $\boldsymbol{\Gamma}$ is positively oriented,

$$\int_{\boldsymbol{\Gamma}} f(z)\,\mathrm{d}z = 2\pi \mathrm{i} \sum_{k=1}^{n} \operatorname*{Res}_{z=z_k} f(z) \quad (\text{p. } 155).$$

C–Riemann conditions If the function $f = u+iv$, where u and v are real-valued functions of z, is differentiable at $z_0 = x_0+iy_0$, then the partial derivatives of u and v at (x_0, y_0) satisfy the equations (known as the *C–Riemann equations*)

$$\left.\frac{\partial u}{\partial x}\right|_{z_0} = \left.\frac{\partial v}{\partial y}\right|_{z_0}, \quad \left.\frac{\partial v}{\partial x}\right|_{z_0} = -\left.\frac{\partial u}{\partial y}\right|_{z_0}.$$

We also have

$$f'(z_0) = \left.\frac{\partial u}{\partial x}\right|_{z_0} + i\left.\frac{\partial v}{\partial x}\right|_{z_0} = \left.\frac{\partial v}{\partial y}\right|_{z_0} - i\left.\frac{\partial u}{\partial y}\right|_{z_0}.$$

(The satisfying of the C–R equations is a necessary but not sufficient condition for f to be differentiable. However, if the four partial derivatives exist and are *continuous* in some neighbourhood of z_0, and satisfy the C–R equations, then f is differentiable at z_0.)

C's root test If Σa_n is a series of positive real terms and

$$\sqrt[n]{a_n} < k < 1 \quad \text{for all} \quad n > N,$$

then Σa_n is convergent. If $\sqrt[n]{a_n} \geqslant 1$ for an infinite number of values of n, then Σa_n is divergent. (*See* the 'limit form' for series with complex terms.)

Limit form: Given a series Σa_n of complex terms, set

$$\rho = \lim_{n\to\infty} \text{lub} \sqrt[n]{|a_n|} \quad \text{(see p. 121).}$$

The series Σa_n is absolutely convergent if $\rho < 1$ and is divergent if $\rho > 1$. If $\rho = 1$, then the test is inconclusive.

C's Theorem If the order of a group G is divisible by a prime p, then G possesses an element of order p (p. 26).

Cayley, Arthur (1821–95)

C group table A method of presenting a finite group of order n by means of a square table of n rows and n columns in which all the possible products of ordered pairs of group elements are listed. (See p. 61 for an example.)

C–Hamilton Theorem Every linear transformation of a vector space to itself (square matrix) is a zero of its characteristic polynomial (p. 83).

C's Theorem Every finite group of order n is isomorphic to a subgroup of the symmetric group (p. 27) of degree n.

Cesàro Ernesto (1859–1906)

C sum p. 142.

Chain rule Let $\mathbf{f}$ be a vector-valued function (p. 126) from E, an open set in $\mathbb{R}^n$, to $\mathbb{R}^m$, and $\mathbf{g}$ be a vector-valued function from an open set in $\mathbb{R}^m$ containing $\mathbf{f}(E)$ to $\mathbb{R}^k$. Assume that $\mathbf{f}$ is differentiable (p. 130) at $\mathbf{a} \in E$, and $\mathbf{g}$ at $\mathbf{f}(\mathbf{a})$. Then the composite function $\mathbf{F} : E \to \mathbb{R}^k$ defined by $\mathbf{F}(\mathbf{x}) = \mathbf{g}(\mathbf{f}(\mathbf{x}))$ is differentiable at $\mathbf{a}$ and

$$\mathrm{d}_{\mathbf{a}}\mathbf{F} = \mathrm{d}_{\mathbf{f}(\mathbf{a})}\mathbf{g} . \mathrm{d}_{\mathbf{a}}\mathbf{f}.$$

(See p. 130. On the right-hand side of the equation we have the product of two linear transformations, represented respectively by a $k \times m$ and an $m \times n$ matrix.)

In terms of partial derivatives we have

$$(\mathrm{D}_j(\pi^i \circ \mathbf{F})|_{\mathbf{a}}) = (\mathrm{D}_l(\pi^i \circ \mathbf{g})|_{\mathbf{f}(\mathbf{a})})\,(D_j(\pi^l \circ \mathbf{f})|_{\mathbf{a}})$$

$(i = 1, \ldots, k; j = 1, \ldots, n; l = 1, \ldots, m)$.

The chain rule is frequently encountered in the special case $k = 1$, i.e. when g and, hence, F are real-valued functions. It is then traditionally written in the form

$$\frac{\partial g}{\partial x_j} = \sum_{l=1}^{m} \frac{\partial g}{\partial \xi_l} \frac{\partial \xi_l}{\partial x_j} \quad (j = 1, \ldots, n)$$

where $\boldsymbol{\xi} = \mathbf{f}(\mathbf{x})$.

In terms of the Jacobian matrix (p. 127) and the gradient (p. 173) the chain rule then becomes

$$\nabla(g \circ \mathbf{f})|_{\mathbf{a}} = \nabla g|_{\mathbf{f}(\mathbf{a})} \,.\, J_{\mathbf{f}}(\mathbf{a}).$$

Change of variable Let $\mathbf{g}$ be a one–one, continuously differentiable (p. 131), vector-valued function from an open set $E \subset \mathbb{R}^n$ to $\mathbb{R}^m$ such that the Jacobian (p. 127) is non-zero on E except for, possibly, a subset of content (p. 160) zero. Let f be a real-valued, continuous function on $\mathbf{g}(E)$. Then for every Jordan-measurable (p. 160), compact (p. 110), subregion S of $\mathbf{g}(E)$,

$$\int_S f(\mathbf{x})\,\mathrm{d}\mathbf{x} = \int_{\mathbf{g}^{-1}(S)} f(\mathbf{g}(\mathbf{t})) \left| |J_{\mathbf{g}}(\mathbf{t})| \right| \mathrm{d}\mathbf{t}.$$

(See example on p. 161.)

(When g is a real-valued function an analogous result is:

Suppose that g has a continuous derivative in $\langle \alpha, \beta \rangle$ and that $g(\alpha) = a$, $g(\beta) = b$. Suppose further that f is continuous in an interval containing $\{g(t) \mid t \in \langle \alpha, \beta \rangle\}$. Then

$$\int_a^b f(x)\,\mathrm{d}x = \int_\alpha^\beta f(g(t)) \frac{\mathrm{d}g}{\mathrm{d}t}\,\mathrm{d}t.$$

The evaluation of integrals by the use of this result is known as **integration by substitution.**)

Choice, Axiom of If $\mathscr{C}$ is a collection of disjoint, non-empty sets S_ν, then there exists a set R which has as its elements exactly one element x_ν from each S_ν.

(See also Zermelo–Fraenkel Axiom VI p. 214.)

(It was shown by Gödel that if the other Z–F Axioms are consistent then so is the system obtained by adding the Axiom of Choice. Cohen later showed that the negation of the Axiom of Choice is also consistent with the other axioms of set theory.)

Cramer, Gabriel (1704–52)
C system p. 48.

D'Alembert, Jean Lerond (1717–83)
D'A's ratio test (i) If $a_{n+1}/a_n \geq 1$ and $a_n > 0$ for all $n > N$ $(N \in \mathbb{N})$, then Σa_n diverges.

(ii) If $a_n > 0$ and there exists $k \in \mathbb{R}$ such that $a_{n+1}/a_n < k < 1$ for all $n > N$, then Σa_n converges. (See limit form for series with complex terms.)

Limit form: Given a series of non-zero complex terms, Σa_n, let

$$r = \lim_{n\to\infty} \text{glb} \left|\frac{a_{n+1}}{a_n}\right|, \quad R = \lim_{n\to\infty} \text{lub} \left|\frac{a_{n+1}}{a_n}\right| \quad \text{(see p. 121).}$$

Then Σa_n is absolutely convergent if $R < 1$ and is divergent if $r > 1$. If $r \leqslant 1 \leqslant R$, the test is inconclusive.

Dedekind, Julius Wilhelm Richard (1831–1916)
D cut p. 65.

De Moivre, Abraham (1667–1754)
De M's Theorem $(\cos\theta + i\sin\theta)^n = \cos n\theta + i\sin n\theta$, $n \in \mathbb{Z}$.

De Morgan, Augustus (1806–71)
De M's Laws $(A \cup B)' = A' \cap B'$, $(A \cap B)' = A' \cup B'$.

Descartes, René (1596–1650)
Cartesian product p. 11.

Diophantos (*c.* 250?)
Diophantine equations p. 46.

Dirichlet, Peter Gustav Lejeune (1805–59)
D function $\phi: \mathbb{R} \to \mathbb{R}$ defined by $\phi(x) = 1$ if $x \in \mathbb{Q}$, $\phi(x) = 0$ if $x \in \mathbb{R} - \mathbb{Q}$.
D's test Let Σa_n be a series of complex terms whose partial sums form a bounded sequence, and (b_n) be a decreasing sequence which tends to 0. Then $\Sigma a_n b_n$ converges.
D's test for uniform convergence Let the sequence of partial sums of the series $\Sigma f_n(z)$ be uniformly bounded (p. 121) on the set E on which each of the f_n is defined. Let (g_n) be a sequence of *real-valued* functions such that $(g_n(z))$ is a decreasing sequence for each $z \in E$ and $g_n \to 0$ uniformly on E. Then $\Sigma f_n(z) g_n(z)$ converges uniformly on E.

Euclid (*c.* 300 B.C.)
E'ean metric p. 99, **plane, space** (p. 165), **vectors** (p. 168).
E'ean algorithm Given $a, b \in \mathbb{Z}$ with $b \neq 0$, then there exist $m, r \in \mathbb{Z}$ such that $a = mb + r$ where $0 \leqslant r < |b|$.
E'ean ring An integral domain R is said to be a Euclidean ring if there is defined a function $d: R - \{0\} \to \mathbb{N}$ satisfying

(1) $\forall a, b \in R - \{0\}$, $d(a) \leqslant d(ab)$;
(2) $\forall a, b \in R - \{0\}$, $\exists t, r \in R$ such that

$$a = tb + r,$$

where either $r = 0$ or $d(r) < d(b)$ (cf. Euclidean algorithm).

Euler, Leonard (1707–83)
E diagram p. 11.
E's constant, γ, is defined to be $\lim_{n\to\infty} \left(\sum_{k=1}^{n} \frac{1}{k} - \log n\right)$.

E's formula $e^{i\theta} = \cos\theta + i\sin\theta$ (p. 165).

E's ϕ (totient) function is defined on the set of positive rational integers by

$\phi(1) = 1$,

$\phi(n)$ for $n > 1$ is the number of positive integers less than n and relatively prime to n.

Fejér, Leopold (Lipót) (1880–1959)

F's Theorem p. 195.

Fermat, Pierre (1601–65)

F's Little Theorem If p is a prime, then $a^p \equiv a(p)$ for all $a \in \mathbb{Z}$.

F's Last 'Theorem' The relation $x^n + y^n = z^n$ cannot be satisfied by positive integers x, y, z and n when $n > 2$.

Fourier, Joseph (1768–1830)

F coefficient p. 193, **series** p. 193.

Fraenkel, Abraham Adolf (1891–1965)

See Zermelo (p. 213).

Frobenius, Georg (1849–1917)

F's Theorem Any division ring (p. 31) which is algebraic over $\mathbb{R}$, the field of real numbers, is isomorphic to either the field of real numbers, the field of complex numbers, or the division ring of quaternions (p. 70).

Galileo, Galilei (1564–1642)

G's paradox A countable set can be put into one–one correspondence with a proper subset of itself.

Galois, Evariste (1811–32)

G group p. 73.

G field A field with p^n elements (p a prime) is known as a Galois field and is often denoted by $GF(p^n)$. (Any two fields of order p^n are isomorphic.)

Gauss, Georg Friedrich (1777–1855)

G'ian field A Gaussian number is a complex number of the form $p + qi$, where $p, q \in \mathbb{Q}$. The Gaussian numbers form a subfield of $\mathbb{C}$, the field of Gaussian numbers. A Gaussian integer is a complex number of the form $m + ni$ where $m, n \in \mathbb{Z}$.

G's Theorem (*Divergence Theorem*) Let V be a region (p. 152) of $\mathbb{R}^3$ which is bounded by the orientable, closed surface S (p. 179). Then, if $\mathbf{f}$ is a continuously differentiable, vector-valued function from V to $\mathbb{R}^3$, $\mathbf{n}$ is the outward-pointing normal to S, and S is 'sufficiently well-behaved',

$$\iiint_V \operatorname{div} \mathbf{f} \, d\mathbf{x} = \iint_S \mathbf{f}.\mathbf{n} \, ds \text{ (pp. 159, 175 and 182).}$$

Gram, Jörgen Pedersen (1850–1916)

G–Schmidt orthogonalisation process Given a basis (p. 40), $\{a_1, \ldots, a_n\}$, of an inner product space V (p. 90) we obtain an orthogonal basis (p. 90) $\{b_1, \ldots, b_n\}$ for V by setting $b_1 = a_1$ and

$$b_m = a_m - \sum_{i=1}^{m-1} \frac{\langle a_m, b_i \rangle}{\|b_i\|^2} b_i \quad (m = 2, 3, \ldots, n).$$

Setting $c_i = b_i/\|b_i\|$, $i = 1, \ldots, n$, we obtain an orthonormal basis for V, $\{c_1, \ldots, c_n\}$.

Green, George (1793–1841)
G's Theorem Let P and Q be two real-valued functions that are continuous on a region $E \subset \mathbb{R}^2$ bounded by a rectifiable Jordan curve $\mathbf{\Gamma}$ and such that D_2P and D_1Q exist and are bounded in the interior of E and that

$$\iint_E D_1 Q \, dx \, dy \quad \text{and} \quad \iint_E D_2 P \, dx \, dy$$

(p. 160) exist.

Then, if $\mathbf{\Gamma}$ is positively oriented, the line integral (p. 177) $\int_{\mathbf{\Gamma}} P\,dx + Q\,dy$ exists and is equal to $\iint_E (D_1Q - D_2P) \, dx \, dy$.

Hamilton, Sir William Rowan (1805–65)
See Cayley (p. 200).
H'ian group A non-abelian subgroup all of whose subgroups are normal. (The Hamiltonian group of smallest order is the quaternion group p. 63.)

Hasse, Helmut (b. 1898)
H diagram p. 75.

Hausdorff, Felix (1868–1942)
H space p. 107.

Heine, Heinrich Eduard (1821–81)
H–Borel Theorem (or *Lebesgue–Borel Theorem*) A set E in $\mathbb{R}^n$ is compact (p. 110) if and only if it is closed and bounded.

H's Theorem If a vector-valued function $\mathbf{f}$ is continuous on a compact subset E of $\mathbb{R}^n$, then it is uniformly continuous on E.

Hermite, Charles (1822–1901)
H'ian adjoint p. 92, **inner product** p. 92, **matrix** p. 92, **transformation** p. 92.

Hilbert, David (1862–1943)
H space p. 112.

Hölder, Otto (1859–1937)
See Jordan (p. 207).

Hospital, Guillaume Francois l' (1661–1704)
l'H's rule Suppose f and g are real-valued functions which are differentiable

in $<a, b>$, that $g'(x) \neq 0$ for all $x \in <a, b>$ (here $-\infty \leqslant a < b \leqslant +\infty$), and that $f'(x)/g'(x) \to A$ as $x \to a$. Then $f(x)/g(x) \to A$ as $x \to a$ provided either

(i) $f(x) \to 0$ and $g(x) \to 0$ as $x \to a$,

or (ii) $g(x) \to \pm\infty$ as $x \to a$.

(A similar rule will hold for limits as $x \to b$.)

Implicit Function Theorem Let $\mathbf{f} = (f_1, \ldots, f_n)$ be a continuously differentiable, vector-valued function mapping an open set $E \subset \mathbb{R}^{n+m}$ into $\mathbb{R}^n$. Let $(\mathbf{a}, \mathbf{b}) = (a_1, \ldots, a_n, b_1, \ldots, b_m)$ be a point in E for which $\mathbf{f}(\mathbf{a}, \mathbf{b}) = \mathbf{0}$ and such that the $n \times n$ determinant

$$|D_j f_i(\mathbf{a}, \mathbf{b})| \neq 0 \quad (i, j = 1, \ldots, n).$$

Then there exists an m-dimensional neighbourhood W of $\mathbf{b}$ and a unique continuously differentiable function $\mathbf{g}: W \to \mathbb{R}^n$ such that $\mathbf{g}(\mathbf{b}) = \mathbf{a}$ and

$$\mathbf{f}(\mathbf{g}(\mathbf{t}), \mathbf{t}) = \mathbf{0} \quad \text{for all} \quad \mathbf{t} \in W.$$

(In the simplest case, when $n = m = 1$, the theorem reduces to: Let F be a continuously differentiable, real-valued function defined on an open set $E \subset \mathbb{R}^2$ and let (x_0, y_0) be a point of E for which $F(x_0, y_0) = 0$ and such that

$$\left.\frac{\partial F}{\partial x}\right|_{(x_0, y_0)} \neq 0.$$

Then there exists an open interval I containing y_0, and a unique function $f: I \to \mathbb{R}$ which is continuously differentiable and such that $f(y_0) = x_0$ and

$$F(f(y), y) = 0 \quad \text{for all} \quad y \in I.)$$

Inverse Function Theorem Let $\mathbf{f}$ be a continuously differentiable, vector-valued function mapping an open set $E \subset \mathbb{R}^n$ to $\mathbb{R}^n$ and let $S = \mathbf{f}(E)$. If, for some point $\mathbf{a} \in E$, the Jacobian (p. 127), $|J_{\mathbf{f}}(\mathbf{a})|$, is non-zero, then there is a uniquely determined function $\mathbf{g}$ and two open sets $X \subset E$ and $Y \subset S$ such that

(i) $\mathbf{a} \in X$, $\mathbf{f}(\mathbf{a}) \in Y$;

(ii) $Y = \mathbf{f}(X)$;

(iii) $\mathbf{f}: X \to Y$ is one–one;

(iv) $\mathbf{g}$ is continuously differentiable on Y and

$$\mathbf{g}(\mathbf{f}(\mathbf{x})) = \mathbf{x} \quad \text{for all} \quad \mathbf{x} \in X.$$

(In the simplest case, when $n = 1$, this theorem becomes: Let f be a continuously differentiable, real-valued function defined on an open interval I. If for some point $a \in I$, $f'(a) \neq 0$, then there is a neighbourhood $<\alpha, \beta>$ of a in which f is strictly monotonic. Then $y \mapsto f^{-1}(y)$ is a continuously differentiable, strictly monotonic function from $<f(\alpha), f(\beta)>$ to $<\alpha, \beta>$. If f is increasing (decreasing) on $<\alpha, \beta>$, then so is f^{-1} on $<f(\alpha), f(\beta)>$.)

Isomorphism Theorems

First Isomorphism Theorem Let $f: G \to H$ be an epimorphism (p. 34) of groups. Then $\operatorname{Ker}(f)$ (p. 35) is a normal subgroup of G and $G/\operatorname{Ker}(f)$ is isomorphic to H.

Second Isomorphism Theorem (*the Diamond Isomorphism Theorem*) If S is any subgroup, and N a normal subgroup of the group G, then $N \cap S$ is a normal subgroup of S and

$$\psi : (N \cap S)s \mapsto Ns$$

defines an isomorphism

$$\psi : S/(N \cap S) \to NS/N$$

(NS is defined on p. 28).

NS
N S
N ∩ S

(If both S and N are normal subgroups of G, then both pairs of opposite sides of the 'diamond' diagram are isomorphic.)

Third Isomorphism Theorem (*the Double Quotient Theorem*) If N is a normal subgroup of G, then every subgroup of G/N has the form R/N for some subgroup R of G. If R is normal in G, then R/N is normal in G/N and

$$\phi : Rg \mapsto (R/N)Ng$$

defines an isomorphism

$$\phi : G/R \to (G/N)/(R/N).$$

Note. The numbering of the isomorphism theorems varies from author to author. Some writers, for example, exclude the First Isomorphism Theorem given above, and, renumbering, add as their Third Isomorphism Theorem what other authors refer to as **Zassenhaus's Lemma**, namely:

If A, B, C and D are subgroups of a group G and if B is a normal subgroup of A, and D a normal subgroup of C, then $B(A \cap D) \lhd B(A \cap C)$ (p. 28), $D(B \cap C) \lhd D(A \cap C)$, and

$$B(A \cap C)/B(A \cap D) \cong D(A \cap C)/D(B \cap C).$$

Jacobi, Carl Gustav Jacob (1804–51)
Jacobian p. 127, **J'ian matrix** p. 127.

Jacobson, Nathan (b. 1910)

J's Theorem Let R be any ring such that for every $a \in R$ there exists a positive integer $n(a) > 1$ satisfying $a^{n(a)} = a$. Then R is a field.

Jordan, Camille (1838–1922)
J arc p. 150, **content** p. 160, **curve** p. 150, **matrix** p. 85, **measurable** p. 160, **region** p. 152.
J Canonical Form Theorem Let V be a finite-dimensional vector space over a field F and t be a linear transformation of V to itself. Then, if the distinct eigenvalues $\lambda_1, \ldots, \lambda_k$ of t are in F, there will exist a basis of V with respect to which the matrix of t is of the form

$$\begin{pmatrix} J_1 & 0 & \ldots & 0 \\ 0 & J_2 & \ldots & 0 \\ & & \ldots & \\ 0 & 0 & \ldots & J_k \end{pmatrix}$$

where each J_i is a reduced (Jordan) matrix (p. 85) in which $\lambda = \lambda_i$.

J Curve Theorem *See* p. 152.
J–Hölder Theorem Any two composition series (p. 61) for a finite group G have the same length and have factors which are isomorphic in some order.

Klein, Christian Felix (1849–1925)
K's four group $D_2 = \{a, b : a^2 = b^2 = (ab)^2 = I\}$.

Kronecker, Leopold (1823–91)
K delta p. 43.

Lagrange, Joseph Louis (1736–1812)
L's identity

$$\left(\sum_1^n a_r^2\right)\left(\sum_1^n b_r^2\right) - \left(\sum_1^n a_r b_r\right)^2 = \sum_{1 \leqslant s < r \leqslant n} (a_r b_s - a_s b_r)^2.$$

L's remainder *See* Taylor (p. 212).
L's Theorem The order (p. 26) of a subgroup of a finite group is a divisor of the order of the group.

Laplace, Pierre Simon (1749–1827)
L's equation p. 176.
L'ian operator p. 176.

Laurent, Matthieu Paul Hermann (1841–1908)
L's Theorem If a complex-valued function is analytic on the *annulus*

$$A(z_0; r_1, r_2) = \{z \mid r_1 < |z - z_0| < r_2\},$$

then it can be represented on A by a series of the form

$$\sum_{-\infty}^{\infty} a_n (z - z_0)^n,$$

where $$a_n = \frac{1}{2\pi i} \int_\Gamma \frac{f(z)}{(z - z_0)^{n+1}} \, dz \quad (n \in \mathbb{Z})$$

and Γ is any positively oriented circle with centre z_0 and radius r where $r_1 < r < r_2$. The series is known as the Laurent expansion (or Laurent series) of f about z_0.

Lebesgue, Henri Léon (1875–1941)
L covering p. 187, **integral** p. 190, **measure** p. 187, **outer measure** p. 187. *See* Heine (p. 204).

Legendre, Adrien Marie (1752–1833)
L function p. 118, **polynomials** p. 193.

Leibniz, Gottfried Wilhelm (1646–1716)
L alternating series test If (a_n) is a (real) sequence which tends monotonically to zero, then the alternating series $\Sigma(-1)^n a_n$ converges.
L formula The nth derivative of the product of two real-valued functions f and g is given by

$$h^{(n)}(x) = \sum_{r=0}^{n} \binom{n}{r} f^{(r)}(x) g^{(n-r)}(x),$$

where
$$h(x) = f(x).g(x) \quad \text{and} \quad \binom{n}{r} = \frac{n!}{r!(n-r)!},$$
whenever the relevant derivatives of f and g exist.

Liouville, Joseph (1809–82)
L's Theorem A complex-valued function which is analytic and bounded everywhere on $\mathbb{R}^2$ is a constant function.

Lipschitz, Rudolf (1832–1903)
L condition A function $f: \mathbb{R} \to \mathbb{R}$ satisfies a Lipschitz condition at x_0 if there exists a constant M and a $\delta > 0$ such that
$$|f(x) - f(x_0)| < M|x - x_0|$$
provided $|x - x_0| < \delta$.

Maclaurin, Colin (1698–1746)
M series Given a real-valued function f with derivatives of all orders in an interval of $\mathbb{R}$ containing the origin, we define the Maclaurin series of f to be
$$\sum_{n=0}^{\infty} \frac{x^n f^{(n)}(0)}{n!} = f(0) + \frac{x}{1!}f'(0) + \frac{x^2}{2!}f''(0) + \ldots + \frac{x^n}{n!}f^{(n)}(0) + \ldots.$$
Provided certain additional conditions are satisfied, for example, that the sequence $(f^{(n)})$ is uniformly bounded (p. 121) on the interval, then the Maclaurin series of f will converge and have the sum $f(x)$.

Mean Value Theorems
First M.V.T. If f is a real-valued function which is continuous on $\leqslant a, b \geqslant$ and differentiable on $<a, b>$, then there is a point $x \in <a, b>$ such that
$$f(b) - f(a) = (b-a)f'(x),$$
M.V.T. for functions of several variables Let $f: \mathbb{R}^n \to \mathbb{R}$ be differentiable (p. 129) at each point of an open set $S \subset \mathbb{R}^n$. If $\mathbf{x}$ and $\mathbf{y}$ are two points of S such that the line segment (p. 150) joining $\mathbf{x}$ to $\mathbf{y}$ lies entirely in S, then there exists a point $\mathbf{z}$ on that line segment such that
$$f(\mathbf{y}) - f(\mathbf{x}) = \nabla f(\mathbf{z}).(\mathbf{y} - \mathbf{x}).$$
M.V.T. for integrals Suppose f is a real-valued function which is continuous on $\leqslant a, b \geqslant$. Then there is at least one point $x \in \leqslant a, b \geqslant$ such that
$$\int_a^b f(t)\,\mathrm{d}t = (b-a)f(x).$$
(More generally, in the case of the Riemann–Stieltjes integral, if, in addition to the above conditions on f, $\alpha : \leqslant a, b \geqslant \to \mathbb{R}$ is a monotonic increasing function, then there exists a point $x \in \leqslant a, b \geqslant$ such that
$$\int_a^b f\,\mathrm{d}\alpha = f(x)[\alpha(b) - \alpha(a)].)$$
See also Taylor (p. 212).

Minkowski, Hermann (1864–1909)
M's inequality

$$\left(\sum_{i=1}^{n}(a_i+b_i)^2\right)^{\frac{1}{2}} \leqslant \left(\sum_{i=1}^{n}a_i^2\right)^{\frac{1}{2}}+\left(\sum_{i=1}^{n}b_i^2\right)^{\frac{1}{2}}.$$

Parseval-Deschênes, Marc Antoine (?–1836)
P's equation p. 195.

Peano, Guiseppe (1858–1932)
P's axioms p. 20.

Pigeon-hole principle (*Kitchen drawer principle*) If n objects are distributed over m places where $m < n$, then there is a place which receives at least two objects.

Principal Axis Theorem If t is a self-adjoint (p. 91) linear transformation of V, a finite-dimensional vector space over $\mathbb{R}$, to itself, then, relative to some orthonormal basis (p. 90) for V, t can be represented by a diagonal matrix whose diagonal entries are the eigenvalues (p. 83) of t appearing with their respective multiplicities.

(*In matrix terms*: Given any real symmetric matrix A, there is an orthogonal matrix P (p. 91) such that PAP^{-1} is a diagonal matrix whose diagonal entries are the eigenvalues of A.)

Pythagoras (*c.* 540 B.C.)
P'ean metric p. 99.

Raabe, Josef Ludwig (1801–59)
R's test If Σa_n is a series of *positive* terms and if

$$n\left(\frac{a_n}{a_{n+1}}-1\right)\rightarrow l \quad \text{as} \quad n\rightarrow\infty,$$

then Σa_n is convergent if $l > 1$ and divergent if $l < 1$.
Ratio test *See* D'Alembert (p. 201)

Riemann, Bernhard (1826–66)
R integral p. 135, **R–Stieltjes integral** p. 137.
See Cauchy (p. 200).

Rolle, Michel (1652–1719)
R's Theorem Suppose that $f: I(\subset \mathbb{R}) \rightarrow \mathbb{R}$ is continuous on $\leqslant a, b \geqslant$ and differentiable on $<a, b>$, and that $f(a) = f(b)$. Then there exists $x \in <a, b>$ such that $f'(x) = 0$.

R's condition A function f is said to satisfy Rolle's conditions on a closed interval $I = \leqslant a, b \geqslant$ if it is continuous on I and continuously differentiable (p. 125) on the open interval $<a, b>$.

Root test *See* Cauchy (p. 200).

Rouché, Eugène (1832–1910)
R's Theorem Let f and g be two functions of a complex variable which are analytic on an open region S (p. 152) of $\mathbb{R}^2$ and let Γ be a rectifiable Jordan curve (p. 151) such that it and its interior lie within S. If $|g(z)| < |f(z)|$ for all z on Γ, then f and $f+g$ have the same number of zeros (p. 155) inside Γ.

Russell, Bertrand Arthur William, 3rd Earl (1872–1970)
R's paradox Let $R = \{x | x \notin x\}$ and consider whether or not R belongs to itself.

Schmidt, Erhard (1876–1959)
See Gram (p. 203).

Schröder, Friedrich Wilhelm Karl Ernst (1841–1902)
S–Bernstein Theorem (*Cantor–Bernstein Theorem*) Given injections

$$u : A \to B \quad \text{and} \quad v : B \to A,$$

then there exists a bijection $w : A \to B$.

Schwarz, Hermann Amandus (1843–1921)
S inequality (*Cauchy–S inequality*) If u and v are any vectors in an inner-product space V (p. 90), then

$$|\langle u, v\rangle| \leqslant \|u\| . \|v\|.$$

Sheffer, Henry Maurice (1883–1964)
S stroke p. 2.

Simpson, Thomas (1710–61)
S's rule Let $f : \langle a, b \rangle \to \mathbb{R}$ be a continuous function and $\{x_0, x_1, \ldots, x_{2N}\}$ a partition (p. 118) of $\langle a, b \rangle$ such that

$$x_k - x_{k-1} = \frac{b-a}{2N} = h \quad \text{say} \quad (k = 1, \ldots, 2N).$$

Then an approximation to $\int_a^b f(t)\,dt$ is given by

$$\tfrac{1}{3}h\{f(x_0) + f(x_{2N}) + 2(f(x_2) + f(x_4) + \ldots$$
$$+ f(x_{2N-2})) + 4(f(x_1) + f(x_3) + \ldots + f(x_{2N-1}))\}.$$

Stieltjes, Thomas Joannes (1856–94)
See Riemann (p. 209).

Stirling, James (1692–1770)
S's formula For large n,

$n!$ is approximately equal to $\surd(2\pi n) . n^n . e^{-n}$,
$\log n!$ is approximately equal to $0.399\,09 + (n + \frac{1}{2}) \log n - 0.434\,294\,5n$.

Stokes, George Gabriel (1819–1903)
S's Theorem Assume that S is a simple parametric surface (p. 179) in $\mathbb{R}^3$ described by a vector-valued function $\mathbf{x}$ defined on a region A of $\mathbb{R}^2$ bounded by a positively-oriented, piecewise smooth, Jordan curve α (p. 150). (The

mixed partial derivatives of $\mathbf{x}$ of the second order (p. 126) are assumed to be continuous on A.) The edge C of S will consequently be described by the composite function $\boldsymbol{\gamma} = \mathbf{x}\boldsymbol{\alpha}$. Stokes's Theorem states that if $\mathbf{f} = (f_1, f_2, f_3)$ is a vector-valued function defined on S for which $\operatorname{curl}\mathbf{f}$ (p. 174), exists, then

$$\iint_S \operatorname{curl}\mathbf{f}.\mathbf{n}\,ds = \int_\gamma \mathbf{f}.d\boldsymbol{\gamma} \quad \text{(pp. 182 and 177).}$$

(In non-vector notation this becomes

$$\iint_S \left(\left(\frac{\partial f_3}{\partial y} - \frac{\partial f_2}{\partial z}\right)\lambda + \left(\frac{\partial f_1}{\partial z} - \frac{\partial f_3}{\partial x}\right)\mu + \left(\frac{\partial f_2}{\partial x} - \frac{\partial f_1}{\partial y}\right)\nu \right) ds$$

$$= \int_C (f_1\,dx + f_2\,dy + f_3\,dz),$$

where λ, μ and ν are the direction cosines of the normal to ds.)

Stone, Marshall Harvey (b. 1903)
S's Theorem Every Boolean algebra is isomorphic to some family of subsets of a set.

Sylow, Ludwig (1832–1918)
S subgroup Let G be a group of order $n = p^r m$ where p is a prime and $p \nmid m$. A p-subgroup (p. 26), $P \subset G$, having order p^r is called a Sylow p-subgroup of G.

S's Theorem Every finite group G contains a Sylow p-subgroup corresponding to each prime p. Any two Sylow p-subgroups are conjugate (p. 58) in G and the number k of distinct Sylow p-subgroups of G is a divisor of the order of G and satisfies $k \equiv 1(p)$.

Sylvester, James Joseph (1814–97)
S's Law The rank and signature (p. 88) of a quadratic form are invariant under change of basis. (In matrix terms: if A is a real symmetric matrix, there is an invertible matrix S such that

$$SAS^T = \begin{pmatrix} I_s & 0 & 0 \\ 0 & -I_{r-s} & 0 \\ 0 & 0 & 0 \end{pmatrix},$$

where r (the rank of A) and s (the signature of A; see the note on p. 88) characterise the congruence class (p. 88) of A.)

Taylor, Brook (1685–1731)
T's Theorem Suppose that $f^{(n-1)}$ is continuous on $\langle a, b \rangle$ and $f^{(n)}$ exists everywhere in $< a, b >$, and that x_0 is any point such that $x_0 \in \langle a, b \rangle$. Then $\forall x \in \langle a, b \rangle$, $x \neq x_0$, there exists a point x_1 which lies inside the interval with endpoints x and x_0 such that

$$f(x) = f(x_0) + \sum_{k=1}^{n-1} \frac{f^{(k)}(x_0)}{k!}(x-x_0)^k + \frac{f^{(n)}(x_1)}{n!}(x-x_0)^n.$$

(The particular result obtained by setting $x = b$ and $x_0 = a$ is also known as the **nth Mean Value Theorem.**)

T series If f has derivatives of every order at each point of an open interval containing x_0, then we define the Taylor series of f about x_0 to be

$$\sum_{n=0}^{\infty} \frac{f^{(n)}(x_0)}{n!}(x-x_0)^n.$$

From Taylor's Theorem it is seen that a necessary and sufficient condition for this series to converge to $f(x)$ is that the '*remainder*'

$$\frac{f^{(n)}(x_1)}{n!}(x-x_0)^n$$

tends to 0 as n tends to ∞. This will certainly be the case if the sequence $(f^{(n)})$ is uniformly bounded on the interval under consideration. (Note that if $x > x_0$, then we have $x_1 = x_0 + \theta(x-x_0)$ where $0 < \theta < 1$.) This form of the remainder is known as the **Lagrange remainder.** Its worth is limited by our lack of knowledge concerning x_1. Alternative forms for this remainder are

$$\frac{(x-x_0)^n}{(n-1)!}(1-\theta)^{n-1} f^{(n)}(x_1),$$

known as the **Cauchy remainder**, and

$$\frac{1}{(n-1)!}\int_0^{x-x_0} (x-x_0-t)^{n-1} f^{(n)}(x_0+t)\,dt,$$

which is known as the **integral remainder** and which reduces to

$$\frac{1}{(n-1)!}\int_0^{x} (x-t)^{n-1} f^{(n)}(t)\,dt$$

in the special case when $x_0 = 0$.

T series for functions of a complex variable If f is analytic on and inside a simple closed curve C and a is a point inside C then the T series of f about a is

$$\sum_{n=0}^{\infty} \frac{f^{(n)}(a)}{n!}(z-a)^n$$

and the series is convergent to $f(z)$ if $|z-a| < \delta$ where δ is the distance from a to the point set C (p. 100).

T series for functions of several variables If $f: A \to \mathbb{R}$, where A is an open set in $\mathbb{R}^m$, is a function having continuous partial derivatives of order n at each point of A, then it is possible to obtain a Taylor expansion for f with an appropriate remainder term. The expansion then takes the form

$$f(\mathbf{a}+\mathbf{x}) = f(\mathbf{a}) + \sum_{i=1}^{m} f_i(\mathbf{a})x_i + \frac{1}{2!}\sum_{i=1}^{m}\sum_{j=1}^{m} f_{ij}(\mathbf{a})x_i x_j + \dots.$$

Unique Factorisation Theorem Every Euclidean ring (p. 202) is a unique factorisation domain (p. 33).

Venn, John (1834–1923)
V diagram p. 11.

Wallis, John (1616–1703)
W's formulae

$$\int_0^{\frac{1}{2}\pi} \sin^{2n} x\,dx = \frac{1.3.\,\ldots\,.(2n-1)}{2.4.\,\ldots\,.2n}\frac{\pi}{2};$$

$$\int_0^{\frac{1}{2}\pi} \sin^{2n+1} x\,dx = \frac{2.4.\,\ldots\,.2n}{3.5.\,\ldots\,.(2n+1)};$$

$$\frac{\pi}{2} = \prod_{n=1}^{\infty} \frac{4n^2}{4n^2-1} = \frac{2}{1}.\frac{2}{3}.\frac{4}{3}.\frac{4}{5}.\frac{6}{5}.\frac{6}{7}.\ldots$$

Wedderburn, Joseph Henry Maclaglan (1882–1948)
W's Theorem A finite division ring is necessarily a field.

Weierstrass, Karl (1815–97)
See Bolzano (p. 198).
W M-test Let (f_n) be a sequence of functions defined on a set $T \subset \mathbb{C}$ and (M_n) a sequence of non-negative numbers such that $|f_n(x)| \leqslant M_n$, for $n = 1, 2, \ldots$ and all $x \in T$. Then $\Sigma f_n(x)$ converges uniformly on T if ΣM_n converges.

Wilson, John (1741–93)
W's Theorem If p is prime, then $(p-1)! \equiv -1(p)$.

Zassenhaus, Hans J. (b. 1912)
See Isomorphism Theorems (p. 206).

Zermelo, Ernst (1871–1953)
Zermelo–Fraenkel Axioms
Undefined terms: sets, the binary relation $\in$ between sets. A domain of sets is postulated and if the relation $a \in A$ holds between the sets a and A, then we say that a is an *element* of A.

Axioms

I. If a, b and A are sets such that $a \in A$ and $a = b$, then $b \in A$.

II. Given sets a, b, then there exists a set denoted by $\{a, b\}$ such that $a \in \{a, b\}$, $b \in \{a, b\}$ and if $x \in \{a, b\}$ then $x = a$ or $x = b$.

III. To every set A which possesses at least one element there corresponds a set $\mathfrak{S}A$ whose elements are all the elements of the elements of A.

IV. To every set A there corresponds a set $\mathscr{P}(A)$, the *power set* of A, whose elements are all the subsets (defined as on p. 8) of A.

V. For every set A and every property $\mathscr{E}$ there is a set whose elements are all those elements of A which possess the property $\mathscr{E}$.

VI (Alternative statement of the Axiom of Choice (p. 201)). If A is a set different from $\varnothing$ (defined as on p. 9), whose elements are disjoint sets (defined as on p. 9), and $\varnothing \notin A$, then there exists a set which is different from $\varnothing$, whose elements are those subsets of $\mathfrak{S}A$ (Axiom III) that have exactly one element in common with each element of A.

VII (Axiom of Infinity – guaranteeing the existence of one infinite set). There exists a set Z such that $\varnothing \in Z$ and, if $x \in Z$, then $\{x\} \in Z$.
Z's Well-ordering Theorem (p. 80).

Zorn, Max August (b. 1906)
Z's Lemma If every simply ordered subset (p. 75) of a poset S (p. 18) has an upper bound, then S has at least one maximal element (p. 75).

Appendix 2 Alphabets used in mathematics

Greek

Α	α	alpha
Β	β	beta
Γ	γ	gamma
Δ	δ	delta
Ε	ϵ	epsilon
Ζ	ζ	zeta
Η	η	eta
Θ	θ	theta
Ι	ι	iota
Κ	κ	kappa
Λ	λ	lambda
Μ	μ	mu
Ν	ν	nu
Ξ	ξ	xi
Ο	o	omicron
Π	π	pi
Ρ	ρ	rho
Σ	σ	sigma
Τ	τ	tau
Υ	υ	upsilon
Φ	ϕ	phi
Χ	χ	chi
Ψ	ψ	psi
Ω	ω	omega

German

𝔄	𝔞	a
𝔅	𝔟	b
ℭ	𝔠	c
𝔇	𝔡	d
𝔈	𝔢	e
𝔉	𝔣	f
𝔊	𝔤	g
ℌ	𝔥	h
ℑ	𝔦	i
𝔍	𝔧	j
𝔎	𝔨	k
𝔏	𝔩	l
𝔐	𝔪	m
𝔑	𝔫	n
𝔒	𝔬	o
𝔓	𝔭	p
𝔔	𝔮	q
ℜ	𝔯	r
𝔖	𝔰 or ſ	s
𝔗	𝔱	t
𝔘	𝔲	u
𝔙	𝔳	v
𝔚	𝔴	w
𝔛	𝔵	x
𝔜	𝔶	y
ℨ	𝔷	z

Index of symbols

Subject index

Where two entries are shown in the index, this indicates that a second definition at a different level has been given. In particular, we indicate that the definition is framed in terms of metric spaces by use of the symbol (M), of topological spaces by (T), of real-valued functions of a real variable by (R), complex-valued functions by (C), of functions of one real variable by (O), and of functions of several real variables by (S).

The reader's attention is drawn to the note printed in bold type on p. 22.